Kira Schumacher, Wolf Fichtner, Frank Schultmann (Eds.)

Innovations for sustainable biomass utilisation in the Upper Rhine Region

PRODUKTION UND ENERGIE

Karlsruher Institut für Technologie (KIT)
Institut für Industriebetriebslehre und Industrielle Produktion
Deutsch-Französisches Institut für Umweltforschung

Band 18

Eine Übersicht aller bisher in dieser Schriftenreihe
erschienenen Bände finden Sie am Ende des Buches.

Innovations for sustainable biomass utilisation in the Upper Rhine Region

Edited by
Kira Schumacher, Wolf Fichtner, Frank Schultmann

Scientific Publishing

This report was elaborated by the project partners of the Interreg IV project 'OUI Biomasse'. The authors of the individual contributions are listed at the beginning of each section and if applicable in the sub-sections.

OUI BIOMASSE

Impressum

KIT Scientific Publishing

Karlsruher Institut für Technologie (KIT)
KIT Scientific Publishing
Straße am Forum 2
D-76131 Karlsruhe

KIT Scientific Publishing is a registered trademark of Karlsruhe Institute of Technology. Reprint using the book cover is not allowed.

www.ksp.kit.edu

Print on Demand 2017 – Gedruckt auf FSC-zertifiziertem Papier

ISSN 2194-2404
ISBN 978-3-7315-0423-8
DOI 10.5445/KSP/1000048433

Preface

Biomass is a renewable resource that can be used as raw material, energy carrier, or feedstock for the production of a range of chemical substances. Because of the limited availability of fossil fuels, biomass demand is expected to rise sharply in the future. This constitutes a major challenge in terms of sustainable development. In particular, the cultivation of energy crops raises questions about land use competition with food and feed crops. Moreover, the energetic use of biomass faces competition from alternative uses by other sectors. Consequently, the use of biomass brings about significant social and environmental challenges, which in some cases result in the rejection of its use by various interest groups.

Therefore, a comprehensive, inter- and transdisciplinary research approach is needed for sustainable biomass utilisation, taking into account the entire value chain, different alternative uses, regulatory frameworks, business environments, as well as locally specific environmental and social conditions. Against this backdrop, the 'OUI Biomasse' project brought together a multidisciplinary project team with scientists from multiple disciplines comprising economists, engineers, forestry scientists, physicists, biologists, chemists, geographers, and sociologists from all major research institutions in the trinational Upper Rhine Region (URR). The cross-border URR served as a particularly suitable study region as it allowed for cross-country comparisons to analyse the impact of differences in national frameworks with regard to biomass use.

The 'OUI Biomasse' project is therefore an excellent example for cross-border, inter- and transdisciplinary research, which has been both a major challenge and achievement of the project. Hence, this report does not only summarise the scientific results of the project but also gives some insights on the encountered challenges regarding cross-border research projects.

Karlsruhe, January 2017

Kira Schumacher, Wolf Fichtner and Frank Schultmann

Acknowledgements

The 'OUI Biomasse' project team thanks the Interreg IV Programme, the Science Offensive of the trinational metropoliton region Upper Rhine (TMO) and all other contributors for this excellent opportunity and the financing of the project. Our thanks are also directed to the associated partners as well as the members of our external Advisory Board for supporting and accompanying closely our work. We also would like to thank the TRION network for energy and climate for the cooperation on communication and event management activities as well the establishment of contacts with experts in the field of bioenergy. Thanks must also go to all participants of the stakeholder workshops conducted in France, Germany and Switzerland as well as all experts in the field who shared with us their valuable knowledge and experience.

Financial support:

Fonds Européen de Développement Régional (FEDER)
Europäischer Fonds für Regionale Entwicklung (EFRE)

Dépasser les frontières : projet après projet
Der Oberrhein wächst zusammen, mit jedem Projekt

interreg Oberrhein Rhin Supérieur

Trinationale Metropolregion Oberrhein – Région Métropolitaine Trinationale du Rhin Supérieur

Wissenschaftsoffensive
L'Offensive Sciences

Wissenschaft Sciences

Baden-Württemberg
Ministerium für Wissenschaft, Forschung und Kunst

Région Alsace

Rheinland-Pfalz
Ministerium des Innern, für Sport und Infrastruktur

Schweizerische Eidgenossenschaft
Confédération suisse
Confederazione Svizzera
Confederaziun svizra

Associated partners:

- Pôle de compétivité Alsace Energivie (now Fibres-Energievie)
- Alsace VITAE
- Ecole Nationale Supérieur d'Architecture de Strasbourg
- European Confederation of Upper Rhine Universities (EUCOR)
- Fraunhofer Institut für System- und Innovationsforschung (ISI)
- Landesforsten Rheinland-Pfalz (Zentralstelle der Forstverwaltung)
- Powerfarm Holding GmbH
- Stadtwerke Karlsruhe GmbH
- TRION

Table of Contents

List of figures

List of tables

1 Background and motivation

***Kira Schumacher**, Wolf Fichtner and Frank Schultmann*

French-German Institute for Environmental Research (DFIU), Karlsruhe Institute of Technology (KIT), Karlsruhe, Germany.

1.1 Motivation and goals

Biomass is a renewable resource that has been used as raw material and energy carrier ever since the beginning of human history. Today biomass is increasingly used to substitute fossil fuels in the energy and transport sector and as material for a range of chemical substances. Due to the limited availability of fossil fuels, political incentives, and changing consumption patterns, biomass demand is expected to rise sharply in the future. The increasing utilisation of biomass, however, comes along with various sustainability challenges (cf. section 4.3). In particular, the cultivation of energy crops raises questions about land use competition with the production of food and feed, the so called "dinner-plate or fuel tank" debate. Other controversial issues associated with the expanding bio-energy sector include potential negative environmental impacts through resource overexploitation and mono-cropping, such as biodiversity losses, soil degradation, as well as air and water pollution. Moreover, biomass use has a significant social dimension making the acceptance of various interest groups an important condition for sustainable biomass use concepts (Kortsch et al. 2015).

Against this backdrop, the 'OUI Biomasse' project was initiated with the vision to establish the tri-national Upper Rhine Region (URR) as one of the most innovative regions in Europe in the field of sustainable biomass utilisation. The URR is a particularly interesting study region as it consists of four sub-regions (Alsace, North-Western Switzerland, Southern Palatinate and Baden) belonging to France, Switzerland and Germany (Statistische Ämter am Oberrhein 2014) and therefore allows for cross-country comparisons. Even though the URR forms a geographically coherent region with regard to natural conditions (e.g. soils, climate), there are substantial differences in legal frameworks, as well as cultures and anthropological views.

To approach the outlined challenges in the URR, the ‘OUI Biomasse’ project aimed at the development of a knowledge-based sustainable biomass strategy for the transition of the regional energy system. In practice, this was done by drafting a “Roadmap for sustainable biomass utilisation in the URR” (OUI Biomasse 2015), to be used as an action plan and a strategic guideline for the implementation of sustainable biomass projects. By involving relevant local stakeholders from politics and the administration, the project aimed to give an important stimulus to environmental policy and innovation for future development of the URR. In addition to these rather strategic goals, the project pursued the following specific objectives:

- study all relevant aspects of the biomass value chain, including the production, the transportation and use of biomass and biomass based products
- take into account the different future development alternatives through scenarios
- analyse potential impacts of possible future developments in terms of sustainability criteria
- launch the local dialogue on the advantages and disadvantages of using biomass with the various stakeholders in the field.

1.2 Inter- and transdisciplinary research approach

To achieve the above defined goals, a comprehensive, inter- and transdisciplinary research approach was required, taking into account the entire value chain, different alternative uses, regulatory frameworks and business conditions, as well as the specific environmental and social situation on a local level.

To account for the multiple dimensions of the biomass topic, the 'OUI Biomasse' project brought together a team of scientists from multiple disciplines comprising economists, engineers, forestry scientists, physicists, biologists, chemists, geographers, and sociologists from all major research institutions in the tri-national URR. Besides multiple viewpoints, approaches and methodologies of the various disciplines, different research cultures and languages needed to be recognized. Moreover, the systemic perspective of the project required a close cooperation and high interaction between the researchers with frequent meetings and intensive discussions. Therefore, the interdisciplinary approach was both a big challenge and important achievement of the project. Some of those challenges connected to the data collection process in the three national sub-regions are described exemplarily in section 3.1.4.

Besides the scientific requirements, the consortium wanted to foster the exchange with relevant actors from outside of the project team in order to create transdisciplinary knowledge and define concrete recommendations for a sustainable development of the region. The cross-cutting tri-national network of scientists therefore needed to interact with a wide variety of other stakeholders, such as political and industrial players, non-governmental organisations as well as the civil society. To enable the participation of external parties, four tri-national workshops and three stakeholder workshops were conducted at different places spread over the whole region. Furthermore, an advisory board was established with experts from science, industry, administration and politics from all three countries.

Through these measures an intensive dialog was held with relevant actors in the three sub-regions throughout the course of the project.

1.3 Research areas

The 'OUI Biomasse' project takes a systemic perspective on bio-energy within the transformation of the regional energy system of the URR. Therefore, the individual research areas (RA) needed to be closely related by several loops, through which data was exchanged and different scenarios were developed and assessed. The thirteen partners from research institutions in France, Germany and Switzerland formed six complementary RAs, as displayed by Figure 1.1. Each of the RAs is briefly described below. For more detailed information on the research approach and results, please refer to the individual chapters of this report.

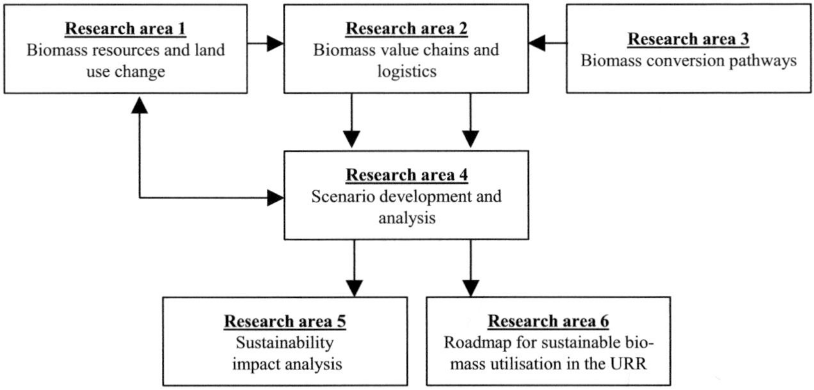

Figure 1.1: Research approach of the 'OUI Biomasse' project

RA 1: Biomass resources and land use change

This first RA was coordinated by the *Professur für Fernerkundung und Landschaftsinformationssysteme* **(FeLis)** from Albert-Ludwigs-University of Freiburg in Germany. The contributing partners include the institutes

Institut für Umweltwissenschaften (**IfU**) of the University Koblenz-Landau, *Laboratoire d'Hydrologie et de Géochimie de Strasbourg* (**LHyGeS**) of the University of Strasbourg/ CNRS, *Laboratoire Image, Ville, Environnement* (**LIVE**) of the University of Strasbourg/ CNRS, *Institute for Technology Assessment and Systems Analysis* (**ITAS**) of the Karlsruhe Institute of Technology and *French-German Institute for Environmental Research* (**DFIU**) of the Karlsruhe Institute of Technology. The task of RA 1 was to identify local biomass resources and land use conflicts by using statistical data, maps, remote sensing and Geographical Information System (GIS) modelling. The ultimate goal was to obtain an inventory of the currently available biomass resources and the land use in the URR (cf. section 3.1). For all three sub-regions, the total agricultural land area and the proportions of the different cultivated crop plants and their respective yields was determined (cf. section 3.1.1). Additionally, forested areas in the URR were identified by using remote sensing imagery and land cover as well as statistic data (cf. section 3.1.2). The amounts of secondary biomass originating from organic household waste, bulk waste, green waste and vineyard residues, which represent a specific fraction of organic waste in the URR, were determined as well (cf. section 3.1.3). Based on the inventory data of the different biomass fractions, technical biomass potentials were estimated, which served as input data for the proceeding RAs, in particular RA 2 and RA 4.

RA 2: Biomass value chains and logistics

The second RA was carried out by the *French-German Institute for Environmental Research* (**DFIU**) of the Karlsruhe Institute of Technology. The task of RA 2 was to develop a planning model for the promotion of cross-border regional production and logistics networks. The major aim was to determine potential production sites for the conversion of biomass into electricity, heat, biogas, or biofuel (cf. section 4.2). The developed economic model allows for an optimisation of the localisation of new bioenergy plants as function of the biomass resources and the costs of investment of the available technologies. RA 2 mainly used data collected by RA 1 and RA 3, linking the identified technical biomass potentials to the existing conversion technologies in order to determine the most suited conversion pathway, location, and capacity of a potential production site. At the centre of these decisions is the trade-off between economies of scale for larger production sites and lower transportation costs for smaller installations. The developed model is able to provide decision support for various levels of the value chain, from the biomass provision, the transport of feedstocks, to the conversion of biomass into the final products.

RA 3: Biomass conversion pathways

The third RA consisted of four sub-working groups, which dealt with biomass conversion from different disciplinary angles. The sub-groups were coordinated by the *French-German Institute for Environmental Research* (**DFIU**) of the Karlsruhe Institute of Technology. Other contributing partners included the institutes *Laboratoire Gestion des Risques et Environnement* (**GRE**) of the University of Upper Alsace, *Institut de Chimie et Procédés pour l'Energie, l'Environnement et la Santé* (**ICPEES**) of the University of Strasbourg, and *Génétique Moléculaire, Génomique, Microbiologie* (**GMGM**) of the University of Strasbourg. The central objective of RA 3 was the techno-economic evaluation of technologies and pathways for providing (electrical and thermal) energy and fuels from biomass (cf. section 3.2.3, Bidart et al. 2014). To this end, existing and new technologies for biomass conversion were technically and economically characterised and assessed with regard to their suitability for the URR. Moreover, a large variety of methodologies was applied to deal with biomass as a feedstock, placing great attention on fundamental issues of its energetic transformation. Here the analysis of conversion pathways for residues from viniculture received special attention. Main processes analysed included methanation (cf. section 3.2.1), combustion (cf. section 3.2.2), and anaerobic digestion (cf. section 3.2.4) of residues from viniculture. The techno-economic data obtained in RA 3 served as input parameters for the economic model developed in RA 2.

RA 4: Scenario development and analysis

RA 4 was carried out by the *Institute for Technology Assessment and Systems Analysis* (**ITAS**) of the Karlsruhe Institute of Technology (KIT). The task of RA 4 consisted in the execution of an exploratory scenario analysis to map the possible future trends until 2030 in cooperation with local interest groups in the URR (cf. section 4). This was done through the identification of main drivers most relevant for the future development of biomass production and use for the URR. The four main drivers determined are 'Energy Policies', 'Agricultural Policies' and 'Nature Conservation Policies' as well as the development of 'Bioenergy Technologies'. The variations of settings of the four main drivers resulted in three scenarios: a 'Business as Usual' scenario (BAU), a 'Maximum Exploitation' scenario (MaxEx) and a 'Conservation and Recreation' scenario (ConsRec). The BAU scenario is mainly based on historical projections and keeping up current trends and framework conditions and serves as reference scenario. The MaxEx scenario is based on political and economic conditions that favour an expansion of the bioenergy sector based on the regional biomass potential. In contrast, under the basic conditions of the ConsRec scenario only parts of the biomass available would be used under stronger ecological guidelines in 2030. The main goal of the scenario analysis is to foster a stakeholder dialogue about possible diverging developments of future biomass production and use in the URR. Results of RA 4 have therefore been presented and discussed during the three stakeholder workshops in Germany, France, and Switzerland. Finally, the scenarios have been contributing substantially to the development of the roadmap elaborated by RA 6.

RA 5: Sustainability impact analysis

RA 5 was coordinated by the *Laboratoire Image, Ville, Environnement* (**LIVE**) of the University of Strasbourg/ CNRS. Contributing partners included the *Association pour la Surveillance de la Pollution Atmosphérique* (**ASPA**), the institutes *Laboratoire d'HYdrologie et de GEochimie de Strasbourg* (**LHyGeS**) of the University of Strasbourg/ CNRS, *Departement Umweltwissenschaften* (**DUW**) of the University of Basel, *Institute for Technology Assessment and Systems Analysis* (**ITAS**) of the Karlsruhe

Institute of Technology (KIT), and *Institut de Chimie et Procédés pour l'Energie, l'Environnement et la Santé* (**ICPEES**) of the University of Strasbourg/ CNRS. In RA 5 an environmental, economic, and social impact analysis of different biomass conversion pathways in the URR was carried out (cf. section 4.3). With regard to environmental impacts, several methodologies have been applied for an in-depth analysis of specific pathways. This included the assessment of impacts on air and soil pollution (cf. section 4.3.1 and 4.3.4) as well as the conduction of life cycle analysis to compare the environmental impact of several biofuels (cf. section 4.3.3). Additionally, soil samples were analysed to assess the effect of the cultivation of miscanthus for bioenergy generation on soil quality (cf. section 4.3.2). These outlined specific studies regarding the environmental dimension have been complemented with considerations about the social and economic impact and incorporated in an integrated sustainability assessment (cf. section 4.3.5). The results of the sustainability assessment were subsequently considered for the development of the roadmap in RA 6.

RA 6: Roadmap for sustainable biomass utilisation in the URR

RA 6 was headed by the *Institut für Ecopreneurship* (**IEC**) of the University of Applied Sciences and Arts Northwestern Switzerland. Other contributing partners included the institutes *Gestion Territoriale de l'Eau et de l'Environnement* (**GESTE**) of the Ecole Nationale du Génie de l'Eau et de l'Environnement de Strasbourg, *Institute for Technology Assessment and Systems Analysis* (**ITAS**) of the Karlsruhe Institute of Technology, and *French-German Institute for Environmental Research* (**DFIU**) of the Karlsruhe Institute of Technology. RA 6 was responsible for the compilation of all results together with recommendations for conversion and business initiatives in the "Roadmap for sustainable biomass utilisation in the URR". The regional roadmap contains information about pathways, options for action and recommendations for political, economic and scientific stakeholders, illustrating possibilities and framework conditions for a sustainable biomass utilisation in the URR. With this roadmap it is intended to facilitate and contribute to a deployment of sustainable energy value chains for biomass in the URR. Moreover, RA 6 carried out an extensive interview study with participants of the bioenergy value chain in the three subregions (cf. section 2.5, Daniel and Bailly 2015, Daniel and Tomson 2015) and conducted three stakeholder workshops in Germany, France, and Switzerland.

1.4 References section 1

Bidart C, McKenna R, Fichtner W 2014, *Fundamentals of Economics for Bioenergy*, Interim report for Task 3.1 of the project 'Biomasse OUI - innovations for sustainable utilization of biomass in the Upper Rhine Region', Karlsruhe: KIT-DFIU, Available from: <http://www.oui-biomasse.info/fileadmin/pdf/Publikationen/Bidart_et_al_Fundamentals_of_Economics_for_Bioenergy__2014_.pdf>. [10 June 2016].

Commission of the European Communities 2009: *The Renewable Energy Progress Report: COM(2009) 192 final*, Brussels.

Daniel, F-J, Bailly, V 2015, *Socio-technical structuring of biogas in France,* Intermediary report for Task 6.2 of the project 'Biomasse OUI - Innovations for sustainable utilization of biomass in the Upper Rhine Region', Strasbourg, ENGEES-GESTE, Available from: <http://www.oui-biomasse.info/fileadmin/pdf/Publikationen/RA62_Biogas_France_EN.pdf>. [10 June 2016].

Daniel, F-J, Tomson M 2015, *Socio-economic structuring of the biomass energy supply in Alsace*, Intermediary report for Task 6.1 of the project 'Biomasse OUI - Innovations for sustainable utilization of biomass in the Upper Rhine Region', Strasbourg: ENGEES-GESTE, Available from: <http://www.oui-biomasse.info/fileadmin/pdf/Publikationen/RA61_Biomass_Alsace_EN.pdf>. [10 June 2016].

Kortsch, T, Hildebrand, J, Schweizer-Ries, P 2015, 'Acceptance of biomass plants - Results of a longitudinal study in the bioenergy-region Altmark', *Renewable Energy,* vol. 83, pp. 690-697.

OUI Biomasse 2015, *Roadmap nachhaltige Biomassenutzung,* Available from: <http://www.oui-biomasse.info/fileadmin/pdf/Publikationen/Roadmap_OUI_DE.pdf>. [10 June 2016].

Statistische Ämter am Oberrhein 2014, *Oberrhein Zahlen und Fakten 2014*, Available from: <http://www.rmtmo.eu/de/services/download.html?file=tl_files/RMT-TMO/oberrhein-zahlen-und-fakten_rhin-superieur-faits-et-chiffres_2014.pdf>. [25.09.2016].

2 Current situation of biomass utilisation in the Upper Rhine Region

Christian Bidart[1], Nadège Blond[2], Christoph Hugi[3], Martin Knapp[4], Rolf Meyer[4], Cyril Pallarès[5], Kira Schumacher[1], Wissal Selmi[2], Jérémy Stark[2], Olga Steiger[3]

[1] French-German Institute for Environmental Research (DFIU), Karlsruhe Institute of Technology (KIT), Karlsruhe, Germany.

[2] Laboratoire Image, Ville, Environnement (LIVE), Université de Strasbourg/ CNRS, Strasbourg, France.

[3] Institute for Ecopreneurship (IEC), University of Applied Sciences and Arts Northwestern Switzerland, Muttenz, Switzerland.

[4] Institute for Technology Assessment and Systems Analysis (ITAS), Karlsruhe Institute of Technology (KIT), Karlsruhe, Germany.

[5] Association pour la Surveillance et l'Etude de la Pollution Atmosphérique (ASPA), Schiltigheim, France.

2.1 Description of the Upper Rhine Region

United Nations rank two Sustainable Development Goals (SDGs) as the most important for the transformational challenge in developed countries. The first is to take urgent action to combat climate change and its impacts and the second is to ensure access to affordable, reliable, sustainable, and modern energy for all (United Nations 2015). A sustainable use of biomass for energy provision can contribute to reach these goals. Although bioenergy contributes only with estimated 2-6% to energy production in Europe and clearly has its limits, it has certain advantages. These are in comparison to fossil energy sources the regional availability of biomass across Europe and the advantage that it can be stored compared to intermittent renewables like photovoltaics or wind energy. From an environmental point of view, bioenergy is renewable and can lower greenhouse gas emissions, but often causes larger impacts than other renewable energies, due to the demand for limited fertile land area, the associated resource requirements to grow, harvest, and convert the biomass, and local emissions especially when burning it. Compared to the specific total energy demand of the Upper Rhine Region (URR) estimated at about 30 000 kWh per person per year, bioenergy covers only a small fraction today and the theoretical potentials seem limited. However, against the background of national goals to increase the production of renewable energy, biomass will have to play a role.

The investigated URR area covers the entire territory of Alsace region (France), the north-west of Switzerland including five cantons as well as a great part of Baden and the extreme south of Rhineland-Palatinate (Germany).

With an area of around 21 500 km², the URR is home to about 6 million inhabitants in 2012 from which 3 million are working (RMT-TMO 2014). There are about 1 776 municipalities and the share of forests and used agriculture area of the total area is 43% and 37%, respectively (Le portail de l'économie alsacienne 2015). These areas are already extensively used as it is shown in section 2.2 "Current land use and biomass production". Most biomass types presented in this section are already used either as material resources or to a certain extent for bioenergy production in several biomass to energy conversion pathways present in the region, which are presented in section 3. Wood is currently the primary bioenergy source mainly for heat production. Agricultural products and residues are used mainly in digestion processes to generate biogas. Digestion is also applied to process organic and green waste. Energy from incineration of waste is also accounted as partly renewable due to the content of organic wastes. The energetic use of biomass causes emissions and other impacts of the environment, which are presented in section 4. The different existing and emerging energy conversion pathways involve many different stakeholders and actors. Key actors and existing value chains are presented in section 2.5. There is a strong political and financial support for bioenergy in all three regions, which is presented in section 2.6.

In the description of the current situation of biomass utilisation the focus of the analysis is on energy conversion pathways and not on material use of biomass. Although material use has not been analysed in detail, it can be stated that in the future the pressure on the energetic use of biomass is expected to increase in favour of material uses. However, the competition between energetic and material use might not be critical in such cases, as many sources of biomass in the URR (manure, municipal solid waste, certain residues from agriculture) currently cannot be processed efficiently in biorefineries.

The 'OUI Biomasse' project was conducted from July 2012 to June 2015. Therefore, most of the research is based on statistical and other data sources from the years 2010 and 2011, if not stated differently. The three investigated parts of the Upper Rhine Region are referred to as France, Germany and Switzerland throughout the section.

2.2 Current land use and biomass production

The land use categories in the URR show a distinctive spatial distribution following the specific topographic structure of the region (Figure 2.1). Overall, around 37% of the URR area is used by agriculture. Arable land is concentrated on the flat of the Rhine valley. Permanent grassland is generally located in the mountainous regions and along the rivers. Viticulture represents only 2% of the total surface, but remains an important economic sector for the URR. The main occurrences of viticulture are on the slopes of the Black Forest, the Vosges and the Kaiserstuhl. Forests cover the highest percentage of the land, with about 43% of the total URR area (European Environmental Agency 2015). They are mainly located in mountainous areas such as Black Forest, Vosges and Jura. Broad-leaved forests are relatively rare in the Black Forest with 10% land cover, but more extensive in the Vosges with 19%. Conifer forests are inversely more important in the Black Forest (18%) than in the Vosges (9%). The main urban agglomerations are Karlsruhe, Strasbourg, Mulhouse, and Basel. The URR has relatively favourable climate conditions. Warm and humid air masses from the Mediterranean coming through the Belfort Gap influence the local climate. In addition, thanks to its distance from the Atlantic, the Rhine Graben situates in a transition zone between oceanic climate and continental. This is characterised by an annual mean temperature around 10°C in most of the Rhine valley. Altogether, this results in favourable conditions for biomass production.

- The major part of the URR area (43% forest area, 37% agricultural area) is already intensively used for biomass production.
- Due to topographic structure and climatic conditions, ORR has favourable conditions for biomass production.

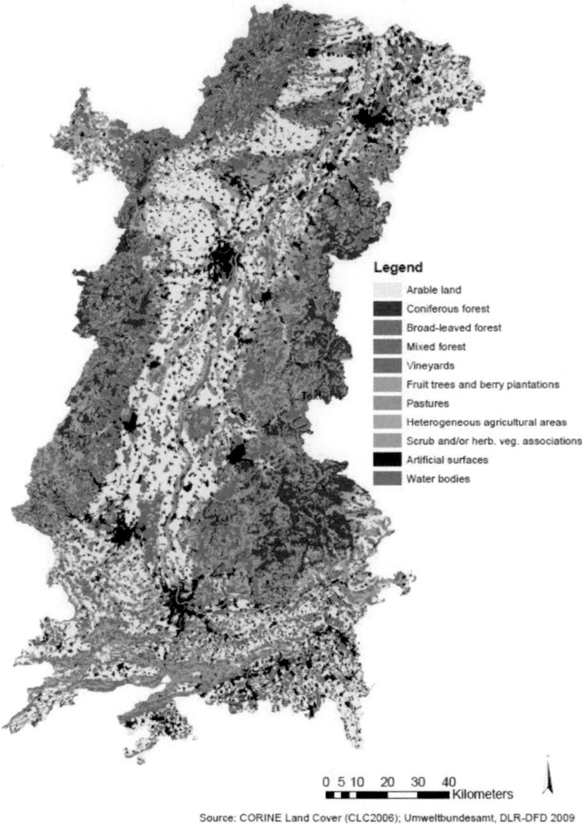

Figure 2.1: Corine land cover map of the URR with relevant land cover classes in 2006 (European Environmental Agency 2015).

Three main sources of biomass have been identified as the most important for energetic use in this study according to their theoretical potential of energy content. Wood as the major bioenergy source mainly used in heat production already contributing about 500 kilowatt-hours per capita per year (kWh/ca./y) in the URR is presented in section 2.2.1. Agricultural residues and manure with a theoretical potential of about 200 kWh/ca./y, but only used to a small percentage, is presented in section 2.2.2. Organic and green waste has a potential of about 90 kWh/ca./y, while sewage sludge from water treatment plants a potential of around 50 kWh/ca./y. These two sources are presented in section 2.2.3.

2.2.1 Forestry biomass

Wood is by far the most important renewable energy source for heat production in the URR. In Figure 2.2 the harvested amounts of stem and industry wood for material use and energy wood are presented. Given the amount of wood from local forests currently dedicated to energy production and the average heat value of wood of 2 650 kWh/m^3, the estimated energy potential from wood is around 500 kWh/ca./y on average in the URR, with 400 in France, 520 in Germany and 570 in Switzerland or about 1-2% of average total end energy consumption per person. The advantages of wood as renewable energy source are its regional availability, the possibility of storage and the production of energy on demand. Estimated amounts for total use of wood per hectare of forest are around 5 m^3 in France, 6 m^3 in Germany and 7 m^3 in Switzerland. Considering very roughly the average growth of wood per hectare to be between 8 and 9 m^3/ha from expert judgement there seem to be a theoretical potential to increase the use of forests slightly in the future.

- Forest biomass for energy production is already extensively used in the URR and an estimated additional potential of around +10% of current exists.
- Main fractions of wood in the URR are harvested as stem wood for material use.
- Energy wood contributes to energy balance with around 400 kWh/ca./y in France, with around 520 kWh/ca./y in Germany and with around 570 kWh/ca./y in Switzerland. This is about 1-2% of total energy demand.

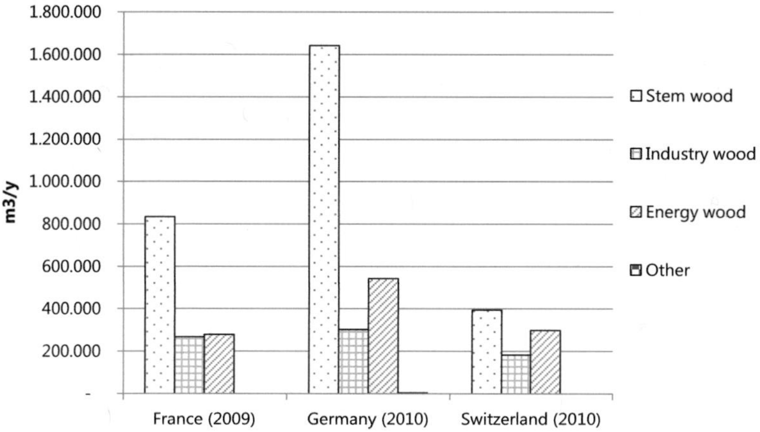

Figure 2.2: Wood harvest and initial purpose of use in France (Agreste Alsace 2011), Germany (ForstBW 2010), and Switzerland (Schweizerische Holzenergiestatistik 2011).

2.2.2 Agricultural land use and biomass production

The total utilised agricultural area (UAA) in the URR is about 783 000 hectare, of which the French part covers 43%, the German part 37% and the Swiss part 20%. The agricultural productions' focus is distinctively different in the three sub-regions (Figure 2.3). Crop production from arable land is dominant in Alsace with around 70% arable land of total French UAA. In Switzerland, the agricultural focus is more on livestock farming, with a high share of permanent grassland, nearly 50% of the total Swiss UAA. In comparison, the lowest share of permanent grassland has Alsace with only around 23%. Permanent cultures such as wine and fruit-growing orchards are quite important in the German URR part (11% of German UAA), also important in Alsace (around 5% of French UAA) and almost negligible in Switzerland. Therewith, agricultural land use differs remarkably between the three sub-regions, causing variant opportunities for bioenergy.

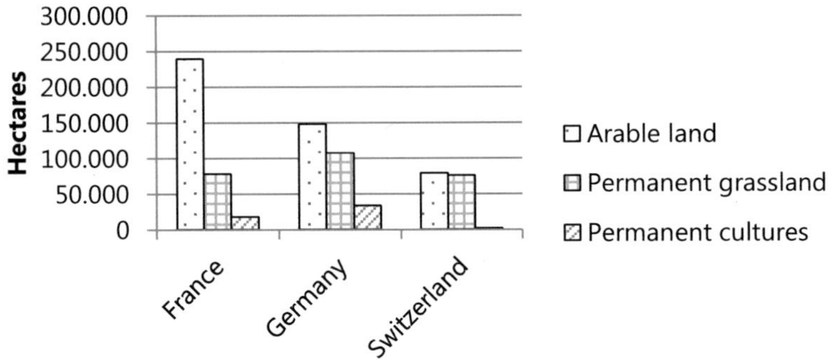

Figure 2.3: Agricultural land use in the URR and the corresponding surfaces in the three countries in 2010 (own calculations based on statistical data).

Overall, the crop production from arable land for food and feed is clearly dominant in the URR. The amount of organic feed and food production (from arable land) is low in all three sub-regions. Energy crops for biogas production are cultivated only in the German sub-region due to the specific feed-in tariffs of the German Renewable Energy Act (EEG) in the past. Production of energy crops for biofuel production is low and exists only in France and Germany due to the EU quota for renewable energy in the transport sector, which has no equivalent in Switzerland. A small part of the animal excrements (slurry, dung) is used for biogas production. Residues from arable crop production (e.g. straw) are currently not used for bioenergy production.

- The three sub-regions show significant differences in agricultural land uses: Alsace has a focus on arable land use, especially corn maize production; Switzerland on permanent grassland and husbandry; the German part of the URR is somewhere in between, with the highest share of permanent cultures.
- Residues from arable crop production (e.g. straw) are currently not use for bioenergy production. However, the energetic content of 50% of agricultural residues generated in the URR is estimated to about 170 kWh/ca./y.
- Assuming that 50% of manure generated would be processed in biogas plants, around 30 kWh/ca./y of bioenergy could be produced in the URR. Currently, only a small part of husbandry manure is used.

2.2.3 Waste, residues and sewage sludge

Under the Waste Framework Directive, the European Union defines waste as "an object the holder discards, intends to discard or is required to discard" (Directive 2008/98/EC). Although waste statistics in the different countries and administrative regions show significant discrepancies in the coding of different waste types, in the following six major waste types with relevance to energetic use of biomass are described.

Household waste is mainly non-recyclable wastes produced by households. A review of household waste analyses from Germany in regions with separate collection of organic waste shows that household waste contains with around 30% of weight a high amount of biomass. The same amount is recorded for Switzerland, but also in regions without separate collection of organic waste (Bundesamt für Umwelt 2014).

Bulk waste is waste, which is too large to be accepted by the regular waste collection. This waste is collected separately or delivered to the waste collection stations. A review of bulk waste analyses from Germany shows that it is composed to around 40% of wood.

Scrap wood is waste wood which has been used as material (e.g. for building or furniture) and which can be used energetically. Estimations of scrap wood generated and used for energy production are difficult to draw for the URR. This is due to intensive imports and exports of stem wood, wood products and scrap wood.

Organic household waste is defined as biodegradable municipal waste deriving from household waste. In some areas of the URR organic waste is collected separately in other regions it is included in household waste. In France and Switzerland separate collection of organic waste is organised on municipal level and is optional. In Germany a recent law stipulates separate collection of organic household waste by 2015. Assumed that 30% of household waste is organic waste, which could be processed with anaerobic digestion, 36 kWh of bioenergy per inhabitant per year could be generated. This estimation does not consider that at the same time the amount of waste incinerated will be reduced.

Green waste stems from private gardens or from municipal organizations responsible for landscaping works and is collected separately in most regions or delivered to the waste collection stations individually. Assumed that all indicated green waste would be digested in biogas plant, the green waste could contribute with 50 kWh/ca./y to bioenergy production.

Sewage sludge from wastewater treatment plants is a waste product generated by municipal and industrial wastewater purification plants. From this sludge, on average, around 30 litres of biogas per inhabitant per day can be generated. Roughly estimated, sewage sludge could contribute with around 50 kWh/ca./y to bioenergy production. Most of the energy produced is being used to cover the energy demand of wastewater treatment plants.

In Figure 2.4, the generated amounts of household waste including bulk waste and green waste are shown per capita for the three regions in the year 2010. Details on computations based on several data sources (for Baden-Württemberg, Ministerium für Umwelt, Klima und Energiewirtschaft; for Rheinland-Pfalz, Statistisches Landesamt; for France, Le Conseil Général du Bas-Rhin and Syndicat Mixte Intercommunal de Traitement des Ordures Ménagères CG68; for Aargau, Statistik Aargau; for Basel Stadt, Präsidialdepartement, Statistisches Amt; Basel-Landschaft: Amt für Umweltschutz und Energie; for Jura, Fistat, Fondattion interjurassienne pour la statistique; for Solothurn, Amt für Umwelt) are given in chapter 3.

- Generated household and bulk waste is currently mainly incinerated. Due to its content of organic material, the energy produced by waste incineration plants is accounted as partly renewable (50%).
- Separate collection of organic fractions of household waste differs significantly between the countries and also between administrative units and municipalities within the countries.

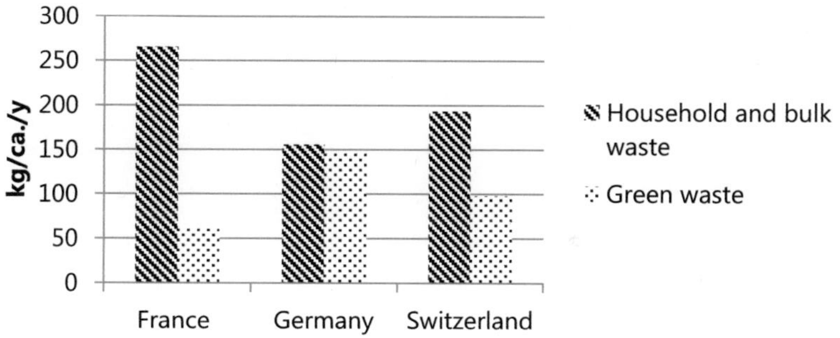

Figure 2.4: Generated household and green waste in kilogram per capita per year by region in 2010 (own calculations based on statistical data).

2.3 Existing plants and technologies

Of existing biomass to energy conversion pathways in the URR three have been identified as major and well established. One is the combustion of woody biomass, second is the anaerobic digestion, which is applied in biogas and sewage gas plants at wastewater treatment plants and third is waste-to-energy, i.e. the incineration of waste. The estimated bioenergy production per capita and year from these pathways in the URR is shown in Figure 2.5.

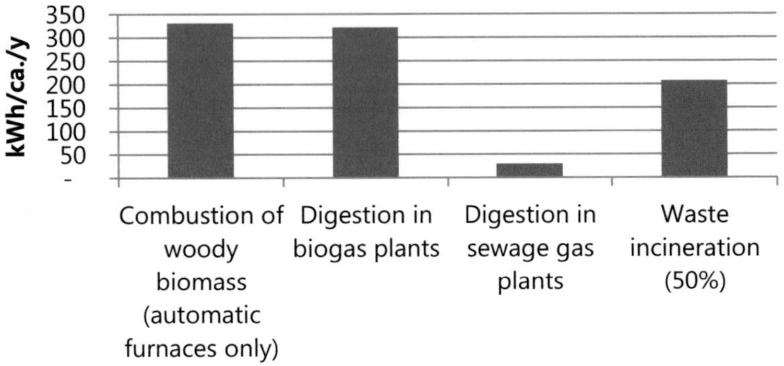

Figure 2.5: Estimated bioenergy production from main biomass to energy conversion pathways in the URR in kWh per capita per year (including 50% of energy from waste incineration plants declared as renewable).

In the following, these pathways and the corresponding plants are presented in more detail.

a. Combustion of woody biomass produces either both heat and electricity by wood heat and power plants or heat only by automatic wood furnaces. Bioenergy is usually generated by using local wood feedstock (logs and chips), imported feedstock (e.g. pellets produced abroad) or residues from wood processing industries. The amount, installed capacity and estimated bioenergy production of automatic wood furnaces in the URR are presented in Table 2.1. There was no data for the UR region of Germany available, so the amount of total wood furnaces in Baden-Württemberg was scaled down to URR.

   For France, no specific information about wood heat and power plants was available. In Germany, there are around 7 wood heat and power plants, with a total installed capacity of around 16 000 kW_{el}. In Switzerland one wood heat and power plant is operated in Basel with a capacity of 37 000 kilowatt-hours electric (kW_{el}) and one additional is in planning. Additionally, there are numerous individual wood furnaces in each country.

Table 2.1: Installed capacity of automatic wood furnaces in France, Germany, and Switzerland.

	France	Germany (est.)	Switzerland
Number of accounted plants in 2011	449	12 500	3 827
Total installed capacity (kW)	148 000	470 000	326 600
Average installed capacity (kW/plant)	330	38	85
Installed capacity per inhabitant (kW/ca.)	0.08	0.17	0.23
Estimated heat production (2 100 h per year) (kWh/y)	310 800 000	987 000 000	685 860 000
Estimated heat production (kWh/ca./y)	167	359	491

b. *Anaerobic digestion* technology is used in *biogas plants* to produce biogas using organic and green waste, other residues and energy crops or in *sewage gas plants* using mainly sewage sludge from wastewater treatment. Generated biogas is mainly converted to electricity and heat, but can also be upgraded to biomethane and fed into the gas grid. However, the production of electricity is more common due to the fact that it is strongly supported by the cost reflective feed-in tariffs. In France, there are five biogas plants situated in Littenheim, Ribeauvillé, Laure, Friesenheim and Obernai. In Germany there are around 74 biogas plants. In Switzerland there are 14 biogas plants processing diverse organic residues from households, industry and agriculture. The amount and capacity of biogas plants in the URR is presented in Table 2.2.

The amount and capacity of sewage gas plants URR is presented in Table 2.3. Most of these plants are producing electricity and heat predominantly for own consumption.

Table 2.2: Energy production of biogas plants in France (Energivie.info 2013), Germany (Landwirtschaft Baden-Württemberg 2015), and Switzerland (Bundesamt für Energie 2013).

	France	Germany (BW)	Switzerland
Number of accounted plants in 2013	5	74	14
Total installed capacity (kW_{el})	2 625	100 837	3 600
Average installed capacity (kW_{el}/plant)	525	1 363	257
Installed capacity per inhabitant (kW_{el}/ca.)	1	37	2.58
Estimated bioenergy production per year (6 000 h/y, electricity production of 1/3 of total) (kilowatt-hours electric and thermal per year (kWh_{el+th}/y))	47 250 000	1 815 066 000	64 800 000
Estimated bioenergy production (kWh_{el+th}/ca./y)	25	660	46

Table 2.3: Installed capacity of biogas plants processing sewage sludge in France (Energivie.info 2013), Germany (Statistisches Amt Baden-Württemberg 2015) and Switzerland (Bundesamt für Energie 2013).

	France	Germany (BW)	Switzerland
Number of accounted plants in 2013	16	45	16
Total installed capacity (kW_{el})	No data	4 365	2 300
Average installed capacity (kW_{el}/plant)	No data	97	144
Installed capacity per inhabitant (W_{el}/ca.)	No data	1.59	1.65
Estimated bioenergy production per year (6 000 h/y, electricity production of 1/3 of total) (kWh_{el+th}/y)	No data	78 576 000	41 400 000
Estimated bioenergy production (kWh_{el+th}/ca./y)	No data	29	30

c. Waste to energy technology produces heat and electricity by combustion of waste. In France, Germany and Switzerland 50% of electricity generated by waste incineration plants is accounted as renewable based on the fraction of organic waste content in municipal waste. In France there are four waste incineration plants, which are situated in Colmar, Strasbourg, Sausheim and Schweighouse sur Moder. In Germany, there is one waste incineration plant situated in Eschbach (Breisgau). Considerable amounts of household waste from German URR are being incinerated in the waste incinerator in Basel and outside the URR as for example in Mannheim. In Switzerland there are five waste incineration plants in the region, the largest one is situated in the city of Basel, one in Solothurn and three in the canton Aargau.

   The amount of waste incineration plants and their energy production is shown in Table 2.4.

Table 2.4: Energy production by waste incineration plants in France (DREAL 2012), Germany (Richers 2010) and Switzerland (own calculation based on reports from individual waste incineration plants). Numbers represent 50% of total production, which is accounted as renewable.

	France		Germany		Switzerland	
Number of plants	4		1		5	
	Thermal	Electric	Thermal	Electric	Thermal	Electric
Production (MWh/y)	237 000	75 500	80 000	60 000	619 500	171 500
Prod. per capita (kWh /ca./y)	168	41	29	22	444	123

Main biomass to energy conversion pathways applied in the URR are wood combustion, anaerobic digestion and waste incineration. These have been widely applied for a long time and considerable expertise has been built up in the region.

2.4 Emissions and impacts from biomass utilisation

The air pollutant sinks and emission sources due to biomass utilisation for bioenergy production and consumption were collected from several institutes in France, Germany and Switzerland. The data and their sources are described in Table 2.5. They were issued from the most recent emission inventories available for the URR in 2015, gridded with the same spatial resolution, but inventories refer to different years (emission inventories need long time to be built and data are not always regularly updated or available).

Table 2.5: Description of the emission inventories collected in the URR.

Sources	Area	Year	Resolution
ASPA (14042904-TD)	Alsace	2010	3km x 3km
LuftHygieneamt Beider	Basel	2000	3km x 3km
Landesanstalt für Umwelt	Baden-Württemberg	2008	3km x 3km
Landesanstalt für Umwelt	Rhineland-Palatinate	2000	3km x 3km

For each region, the data detail the emissions of sulphur dioxide (SO_2), nitrogen oxides (NO_x), carbon monoxide (CO), non-methane volatile organic compounds (NMVOC), methane (CH_4), ammonia (NH_3), particles (PM2.5, PM10, and total PM including particles over 10µm) per activity sector (EMEP/CORINAIR, 2011). For some regions only, emission data were available for benzene (BENZ), styrene (STYR), toluene (TOL) and polycyclic aromatic hydrocarbon (PAHs, benzo[a]pyrene BAP, benzo[a]anthracene BAA, benzo[b]fluoranthene BBF, benzo[j]fluoranthene BJF, benzo[k]-fluoranthene BKF).

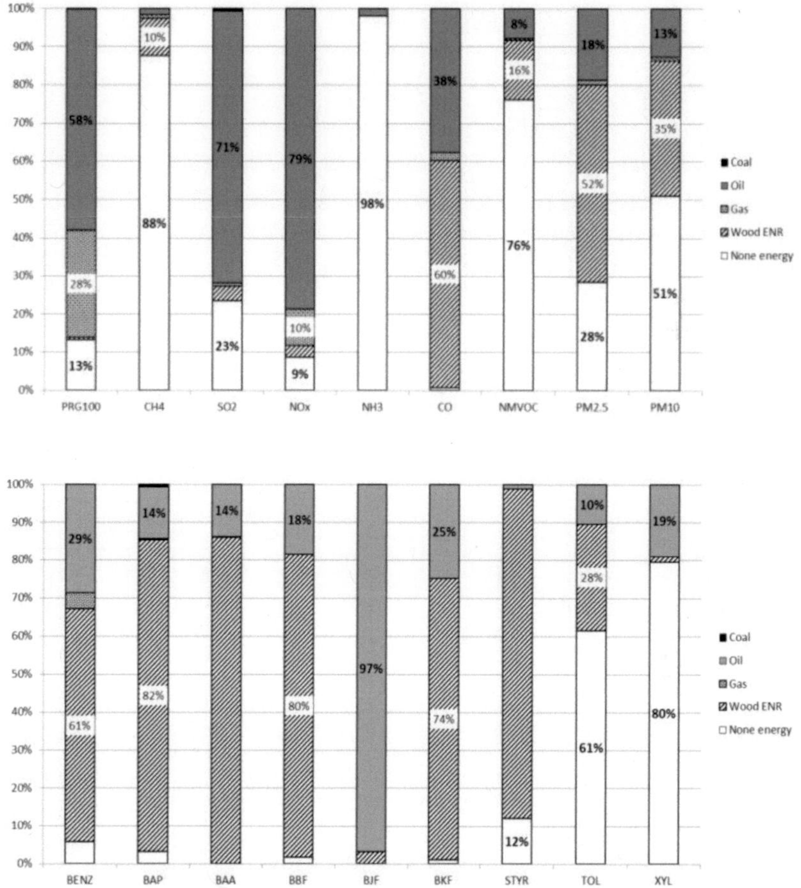

Figure 2.6: Contribution of energy sources in total air pollutant emissions (in %) in the Alsace Region (France, 2010).

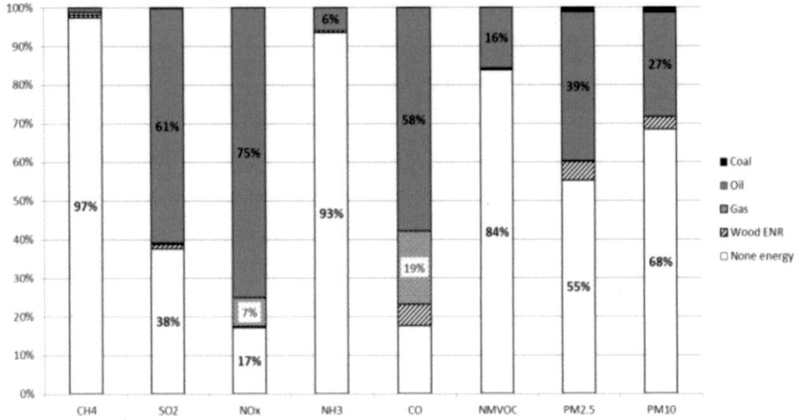

Figure 2.7: Contribution of energy sources in total air pollutant emissions (in %) in the Basel Region (Switzerland, 2000).

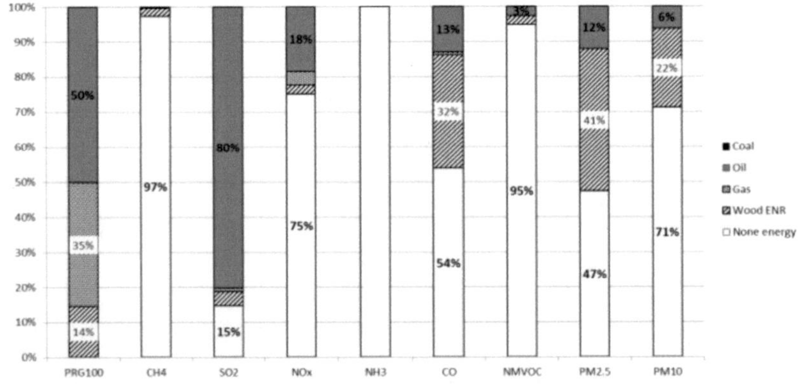

Figure 2.8: Contribution of energy sources in total air pollutant emissions (in %) in Baden-Württemberg (Germany, 2008).

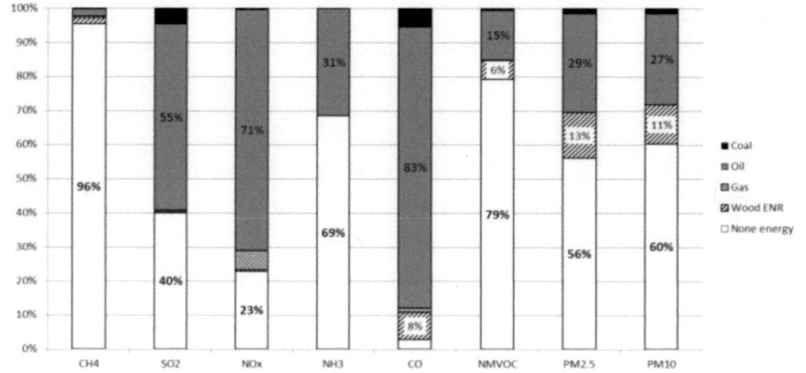

Figure 2.9: Contribution of energy sources in total air pollutant emissions (in %) in Rhineland-Palatinate (Germany, 2000).

When the emission inventories also include greenhouse gases with anthropic (CO_2) and biogenic (CO_2 BIO) carbon dioxide and nitrous acid (N_2O), total emissions of CO_2, CH_4, and N_2O are reported in tons equivalent of CO_2 into a global index PRG100 using their respective global warming potentials.

Figure 2.6, Figure 2.7, Figure 2.8, and Figure 2.9 show the contribution of each type of energy sources in the total emissions for each air pollutant and each region.

Since the emission inventories did not concern the same period of time and are not built using the same methodology, the comparison remains difficult, especially when focusing on biomass utilisation with fast increases in each country. Nevertheless, one can note that oil and gas utilisation contributes to a large amount of air pollutant emissions and is the main source of GES, SO_2, NOx, BJF. It is also shown that the use of biomass is now responsible of large parts of CO, PM, HAPs, benzene, styrene and toluene emissions.

The Alsace Region and Baden-Württemberg show similar patterns except for NOx and CO. For Baden-Württemberg the part classified as "None energy" is much larger for those pollutants especially. "None energy" is associated in France to emissions from activities not directly related to the consumption of energy, like for example particle emissions from land-use in agriculture sector, building construction or abrasion of wheels, brakes, etc.). It may include in some other cases the emissions for which energy source was not detailed.

The emissions due to the use of wood biomass energy represent 19 % of the total PM emissions with 3.49 tons of PM in 2010 at the scale of the URR. The particles over 10 µm are largely originated from non-energetic sectors like agriculture, chemical industry and road transport.

The emissions of PM2.5 and the contribution of wood biomass appear to be more important in France than in other surrounding countries. The contribution of wood biomass reaches 51% of PM2.5 emissions in France in 2010 and 41% in Baden-Württemberg in 2008. The differences can be explained by the more important utilisation of efficient and controlled individual wood heating systems in Germany, compared to France.

It is also important to notice that the emissions over Rhineland Palatinate and Basel Regions collected (reference year 2000) are probably too old and do not take recent energy transition effects into account to be here useful for comparison.

Wood biomass utilisation has an important contribution to local URR air pollution, and especially for the French part of the URR .

2.5 Stakeholders, key actors and existing value chains

There are numerous players in the field of bioenergy in the URR which often resume more than one role depending on the specific set-up of the project. Some of the players from different fields of activities were interviewed in the course of the project: planning and construction companies for bioenergy plants, bioenergy plant owners and operators, biomass and technology provider, regional institutions and associations (Figure 2.10).

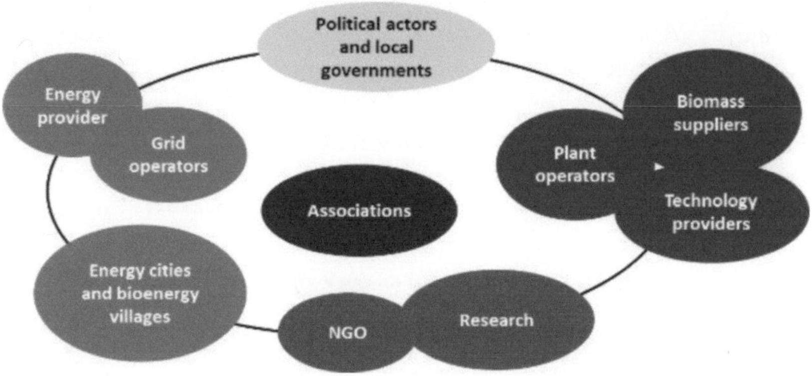

Figure 2.10: Main actors active in the field of bioenergy in the URR.

On national level, different players are often connected through business relations (e.g. contracts), but also through formal and informal networks. Important contracts between players in the field of bioenergy are these between biomass providers and bioenergy plant operators. Players are also organised in several associations, which foster the exchange of information and knowledge. Overall, all players are influenced by national and regional institutional and legal frameworks, which strongly influence the development of bioenergy projects. Based on national and regional legislation, the main influencing factors are financial support schemes, such as feed-in tariffs systems for electricity, other subsidies and technical regulations. Relatively few players are active in more than one country of the Upper Rhine region. These are mainly planning and construction companies and technology providers. The latter are situated mostly outside the URR. Regional institutions and association in the field of biomass and bioenergy have rather informal contacts for information and knowledge exchange.

2.6 Political and legal framework for bioenergy production

A similarity in legal frameworks and strategies of the three countries is that there is strong political support for bioenergy production, in form of strategies and incentives, which support the use of biomass as renewable energy sources. Therefore, all three countries introduced cost-reflective feed-in tariffs for electricity produced from biomass. For heat production from wood, there are also supportive financial measures in place. Differences in the legal framework and strategies considering biomass use for energy production are mainly on the use of energy crops for biogas production, wood use for heat production and biofuel quota. While in Switzerland the use hierarchy of "food, feed, tank" is applied very strictly, in Germany the legislative and financial support for energy crops production was restricted only in 2014, so that new biogas installations will only receive financial support if they use waste and residues in the future. Energy crops will nevertheless still be used in Germany and France to produce biofuels for which European and national quota exists, which set a minimum addition of biofuels to conventional fuels. Sustainability issues in biofuel production are considered in all three countries. However, Switzerland has no national quota for biofuels, but further reaching specifications for comprehensive environmental and social criteria, which have to meet in addition to positive energy balance.

Biomass is the main renewable energy source in France (second in Alsace region after hydraulic energy). The bioenergy production from biomass is mainly driven by the need to fulfil the objectives set by the European Union through the Directive 2009/28/EC, which stated that 20% of EU final energy consumption has to be covered by renewable sources by 2020. This target was increased to 23% in France in general and additional sectoral targets were formulated, 33% for heat, 27% for electricity and 10% for transportation. A detailed national action plan (Plan d'action national en faveur des énergies renouvelables) was proposed to the European Commission in 2010. In the "Grenelle 1" law (2009) a global legislative framework was given and in "Grenelle 2" (2010) specific objectives per sector were defined. The laws were declined at regional scale through the regional plans for climate, air, and energy (Schéma Régional Climat-Air-Energie; DREAL 2012), and urban plans for energy and climate (Plan Climat Energie Territorial PCET) for urban areas with more than 50 000 inhabitants. This general legislative framework will be soon affected by legislative changes on energetic transition for green growth, adopted by the Senat in March 2015 and still in discussion at the National Assembly. The developments are also supported by several specific regulations on renewable electricity: Loi sur la modernisation et le développement du service public de l'électricité (2000), Loi Nouvelle Organisation du Marché de l'Électricité (2010), Arrêté fixant les conditions d'achat de l'électricité issue de l'énergie solaire (2011). Regulations on renewable heat also help to promote the use of renewable energy, as for example Réglementation thermique (2012). Other financial support such as «Fonds Chaleur» and «Appel à projets Biomasse Chaleur Industrie Agriculture Tertiaire» helps to support bioenergy production. The production of biofuels is mainly developed thanks to regulations aiming at reducing the use of fossil fuels: for example the tax named "Taxe Intérieure de Consommation sur les Produits Energétiques" with reduced taxes for biofuels compared to fossil fuels (2011). National regulations are then locally supported through regional calls and associated financial incentives.

In Germany, main regulations and strategies considering biomass use and bioenergy production are the following.

On energy and biomass in general

- Erneuerbare Energie Richtlinie 2009/28/EC
- Energiekonzept 2050 der Bundesrepublik Deutschland (2010)
- Nationaler Biomasseaktionsplan (2010)
- Gesetz zur Förderung des Klimaschutzes in Baden-Württemberg (KSG BW) (2013)
- Integriertes Energie- und Klimaschutz Konzept (IEKK) Baden-Württemberg (2014)
- Landesklimaschutzgesetz (LKSG) Rheinland-Pfalz (2014)

- Biomasseverordnung
- Biomassestrom-Nachhaltigkeitsverordnung

Specific regulations on renewable power are

- Erneuerbare-Energien-Gesetz (EEG) (Revision 2014)
- Kraft-Wärme-Kopplungsgesetz (KWKG) (2002)

Specific regulations on renewable heat are

- Erneuerbare-Wärme-Gesetz (EEWärmeG) (2011)
- Erneuerbare-Wärme-Gesetz Baden-Württemberg (EWärmeG) (2010)

Production of biofuels is regulated according to

- Gesetz zur Änderung der Förderung von Biokraftstoffen (BioKraftFÄndG) (2009)

The basis for biomass use in Switzerland is built on two national strategies, "Biomassestrategie Schweiz" from the year 2009 and "Biomasse-Energiestrategie Schweiz" from the year 2010. These two strategies postulate first of all the preference order of "food, feed, tank" which leads to the fact that biogas and biofuels are produced exclusively from waste and residues in Switzerland. The strategic goals for *renewable energy* are formulated in the national "Energiestrategie 2050" from 2013. In this document, national goals for future bioenergy production are postulated, which foresee an increased amount of bioenergy production. The implementation of these goals will have to be elaborated on cantonal level and is a process which in now about to start. However, three of five northwestern cantons, Baselland, Aargau and Solothurn have recently formulated their own cantonal energy strategies. Legislative framework for *renewable power* in Switzerland is the "Stromversorgungsgesetz" (StromVG) from the year 2007 and "Energiegesetz" (EnG) from the year 1998 which have been amended in 2011. In this framework the cost-reflective feed-in tariffs for renewable electricity are regulated. *Renewable heat* is not regulated on national, but only on cantonal and municipal level, mainly through "Mustervorschriften der Kantone im Energiebereich" (MuKEn). The position paper "Positions-papier biogene Treibstoffe" (2008) sets several requirements which have to be met for *biofuels production*. Although biofuels are not directly promoted in Switzerland, they are excluded from the mineral fuels taxes according to "Mineralölsteuerverordnung" (MinöStV) (2008). As energy provision from agricultural products is not supported in Switzerland, especially the biogas production is based on *waste and residues* such as organic household and industrial waste, green waste and sludge form wastewater treatment plant. The regulation of the waste streams is done by the "Technische Verordnung über Abfälle (TVA)" which is currently in the process of total revision.

- Political support for bioenergy production is present in all three regions, although incentives for bioenergy production differ between the regions.
- Due to growing knowledge about environmental and social impacts associated with bioenergy production, more restrictive concepts are applied recently which differ between the regions in their extent.

2.7 References section 2

Agreste Alsace 2011 : exploitations forestières et scieries d'envergure pour une forêt d'exception, N°12 juillet 2011, Available from: http://agreste.agriculture.gouv.fr/IMG/pdf_R4211A14.pdf. [16 May 2016].

DREAL 2012, Schéma régional Climat Air Énergie Alsace (Document intégral), L'incinération des déchets (p. 155). Available from: <http://www.energivie.info/sites/default/files/uploads/strategie_regionale/scrae/3-rapport.pdf>. [16 May 2016].

EMEP/CORINAIR 2011, Emission Inventory Guidebook, European Environment Agency: Classification of the polluting activity with the SNAP nomenclature and the fuel classification NAPFUEL. Available from: <http://www.eea.europa.eu/publications/technical_report_2001_3>. [16 May 2016].

European Environmental Agency 2015, *Corine land cover 2006 (CLC2006) 100 m - version 12/2009*, Available from: <http://www.eea.europa.eu/data-and-maps/data/corine-land-cover-2006-clc2006-100-m-version-12-2009>. [16 May 2016].

ForstBW 2010, *Jahresbericht 2010 des Landesbetriebes ForstBW. 56. Jahrgang, Forstwirtschaftsjahr 2010,* Available from: <http://forstbw.de/fileadmin/Website_pictures/Gescha%CC%88ftsberichte/ForstBW_ ; JahresberichtMaterialband_2010.pdf>. [16 May 2016].

Le portail de l'économie alsacienne 2015, *Mémento de la statistique agricole 2012*, Available from: www.alsaeco.com. [16 May 2016]

Energivie.info 2013, *Rapport final sur etat des lieux des gisements et de la gestion de la matière organique en Alsace, perspectives de développement des installations de production de biogaz,* Available from: <http://www.energivie.info/sites/default/files/documents/rapport-final-biogaz.pdf>. [16 May 2016].

Energivie.info 2011, *Base de données chaufferies bois,* Available from: http://www.energivie.info/sites/default/files/uploads/decouvrir_energies_renouvelables/bois_collectivites/14-05-12-energivie-base-donnees-chaufferies-bois-01-12-2011.pdf. [16 May 2016].

Richers, U 2010, *Abfallverbrennung in Deutschland – Entwicklungen und Kapazitäten*, KIT Scientific Publishing 2010, Available from: <http://www.itas.kit.edu/pub/v/2010/rich10a.pdf>. [16 May 2016].

Région Metropolitaine Trinatioanle du Rhin Superieur (RMT), Trinationale Metropolregion Oberrhein (TMO) 2014, *Oberrhein Zahlen und Fakten – Rhin Supérieur Faits et Chiffres 2014*, Available from: <http://www.rmtmo.eu/de/services/download.html?file=tl_files/RMT-TMO/oberrhein-zahlen-und-fakten_rhin-superieur-faits-et-chiffres_2014.pdf>. [16.05.2016].

Statistisches Amt Baden-Württemberg 2015, *Klärgasgewinnung und Stromerzeugung am Standort der Kläranlagen in Baden-Württemberg seit*

2008 nach Kreisen, Jahr 2013, Available from: <http://www.statistik.baden-wuerttemberg.de/UmweltVerkehr/Landesdaten/EN_Klaergas_KR_0000.asp>. [May 2015].

United Nations 2015, *Sustainable Development Goals*, Avaiable from: <https://sustainabledevelopment.un.org/sdgs>. [16 May 2016].

Untere Landwirtschaftsbehörden 2004-2014, *Entwicklung der Biogas-anlagen in Baden-Württemberg (Stand 31.12.2014)*, Available from: <https://www.landwirtschaft-bw.info/pb/MLR.Landwirtschaft,Lde/Startseite/Erneuerbare+Energien/Biogas>. [May 2015].

3 Collection of basic data on land use, biomass potentials, bioenergy technologies and plant types

Rahim Aguejdad[1], Ludovic Besaury[2], Jean-Francois Brilhac[3], Alain Brillard[3], Claire Courson[4], Wolf Fichtner[5], Andreas Fritz[6], Christian Bidart[5], Nadège Blond[1], David Elsässer[7], Christelle Gruffaz[2], Muhammad Farhan Ul Haque[2], Martin Knapp[8], Barbara Koch[6], Russell McKenna[5], Rolf Meyer[8], Ann-Kathrin Müller[5], Dong-Binh Nguyen[1], Alexandra Pehle[8], Anne-Cécile Roger[4], Aziz Serradj[1], Grzegorz Skupinski[1], Cornelius Schönnenbeck[3], Gwenaëlle Trouvé[3], Audrey Waldvogel[4], Christiane Weber[1], Maximilien Valente[3], Stéphane Vuilleumier[2]

[1] Laboratoire Image, Ville, Environnement (LIVE), Université de Strasbourg/ CNRS, Strasbourg, France.

[2] Génétique Moléculaire, Génomique, Microbiologie (GMGM), Université de Strasbourg, Strasbourg, France.

[3] Laboratoire de Gestion des Risques et Environnement (GRE), Université de Haute Alsace, Mulhouse, France.

[4] Laboratoire des Matériaux, Surfaces et Procédés pour la Catalyse (LMSPC), Université de Strasbourg, Strasbourg, France.

[5] French-German Institute for Environmental Research (DFIU), Karlsruhe Institute of Technology (KIT), Karlsruhe, Germany.

[6] Professur für Fernerkundung und Landschaftsinformationssysteme (FeLis), Albert-Ludwigs-Universität Freiburg, Freiburg; Germany.

[7] Institut für Umweltwissenschaften (IfU), Universität Koblenz-Landau, Landau, Germany.

[8] Institute for Technology Assessment and Systems Analysis (ITAS), Karlsruhe Institute of Technology (KIT), Karlsruhe, Germany.

3.1 Collection of data on biomass potentials

***Andreas Fritz*[1]*, Christian Bidart*[2]*, Nadège Blond*[3]*, David Elsässer*[4]*, Martin Knapp*[5]*, Barbara Koch*[1]*, Rolf Meyer*[5]*, Ann-Kathrin Müller*[2]*, Alexandra Pehle*[5]*, Rahim Aguejdad*[3]*, Aziz Serradj*[3]*, Grzegorz Skupinski*[3]*, Dong-Binh Nguyen*[3]*, Christiane Weber*[3]

[1] Professur für Fernerkundung und Landschaftsinformationssysteme (FeLis), Albert-Ludwigs-Universität Freiburg, Freiburg; Germany.

[2] French-German Institute for Environmental Research (DFIU), Karlsruhe Institute of Technology (KIT), Karlsruhe, Germany.

[3] Laboratoire Image, Ville, Environnement (LIVE), Université de Strasbourg/ CNRS, Strasbourg, France.

[4] Institut für Umweltwissenschaften (IfU), Universität Koblenz-Landau, Landau, Germany.

[5] Institute for Technology Assessment and Systems Analysis (ITAS), Karlsruhe Institute of Technology (KIT), Karlsruhe, Germany.

In the 'OUI Biomasse' project, the current biomass production and utilisation (section 2) as well as future biomass scenarios (section 4) are analysed at the whole URR scale. All these analyses are based on various spatialized data. In this section we describe the processes of data collection, assessment, processing and exploitation regarding for the project relevant biomass categories *agriculture*, *forest* and *waste*. At this point, it is obvious, that the data collection plays a major role for the project, since it is the basis for all processes within the different facets of OUI Biomasse. Particularly, with regard to the biomass scenarios elaborated in section 4.1 and the identification of potential production sites (section 4.2) the collection and production of spatialized data becomes vital.

In general, data collection for the spatial dimension (*and* semantic dimension) of a metropolitan becomes very challenging. Choosing a common reference date, defining a coherent spatial unit and coming to an agreement regarding definitions are just a few crucial considerations, which were made in this part of the project. We are addressing these issues therefore in an extra sub section 3.1.4, where we discuss in depth the pitfalls and possible solutions to the problems we were facing.

3.1.1 Approach to assess agricultural land use and biomass production data

The modelling of agricultural activities is mainly based on farm structure survey data. Spatial resolution is as far as possible the district level (Land-/Stadtkreis in Germany, Arrondissement in France, Kanton in Switzerland). The reference year 2010 represents the current situation.

3.1.1.1 Agricultural land use

Starting point is the total agricultural land use (Utilised agricultural area – UAA) and the total areas of the three main agricultural land use categories arable land, permanent grassland and permanent cultures (BFS 2013a + b + c, DRAAF 2012a; SL BW 2011a; SL RLP 2012a). The next step is the assessment of agricultural land use by organic farming. The UAA of organic farming is extracted from BFS 2014, Insee 2014b, SL BW 2014a and SL RLP 2012b. For Alsace, statistical data on organic UAA is only available at the level of département. Classes of organic UAA for each commune are provided by OPABA (2013, p. 5) and used to calculate the organic UAA for the arrondissement. Data on main organic agricultural land use categories are directly available from the statistical database for Rheinland-Pfalz (SL RLP 2012b) and Switzerland (BFS 2014a). For the Baden-Württemberg part of the URR, data from the evaluation of the MEKA programme (LEL 2011) are used and extrapolated to the overall organic UAA. For Alsace, the distribution of total areas of arable land, permanent grassland and vineyard area at arrondissement level are also applied for the distribution of these categories in organic farming due to missing data for organic farming. Following this, the UAA and the area of arable land, permanent grassland and permanent cultures under conventional farming are calculated by subtracting organic areas from total areas.

Arable land use is further differentiated in eight main arable land use categories with overall 20 cropping systems (data sources Agreste 2011, BFS 2014a + b, DRAAF 2012a + b, Insee 2014b, SL BW 2011b, SL RLP 2012a):

- Cereals: Wheat, rye, barley, oat, grain maize, other cereals
- Forage crops: Silage maize, cereals as total crops, legumes, fodder beet, temporary grassland
- Root crops: Potatoes, sugar beet
- Oil crops: Rapeseed, sunflower, soy
- Other commercial crops
- Legumes
- Vegetables/horticulture
- Fellow land

For the French sub-region, data at arrondissement level are only available for total cereals, wheat, corn maize, total forage crops and silage maize, for the other crops data are at département and/or region level. In the case of areas of cereals without arrondissement data, the areas at départment level are distributed to the arrondissement.

The total areas of arable crops are in the next step allocated to organic and conventional farming. Statistical data for both farming systems are directly available for the Swiss Kantons. For Baden-Württemberg, Rhineland-Palatinate and Alsace, data on organic arable crop cultivation are available only at a lower level of regional and/or product differentiation. For Baden-Württemberg, the area of the main organic arable crop categories is based on the MEKA evaluation (LEL 2011). Data at district level are not available for Rhineland-Palatinate, only information on the organic cultivation areas of the main crops for the overall federal state (MWVLW RP 2008). It is assumed that this distribution of organic crop areas applies also for the three URR districts. This distribution was also used for the calculation of organic cereal crop areas at district level in Baden-Württemberg. For Alsace, data at département level on the area of the organic main arable crop categories and cereal crops exist (Insee 2014, OPABA 2010), and the cereal départment areas are distributed to the arrondissement. The conventional arable cropping areas are once again calculated by subtracting organic areas from total areas.

In an additional step, (conventional) arable land for food and feed production and for bioenergy production is separated by assessing arable land cultivated with energy crops. Part of the biomass from energy crops is locally/regionally converted into energy (e.g., in biogas plants). Another part is traded nationally and/or internationally (e.g., biomass of 1st generation biofuel crops) so that the bioenergy production is not linked necessarily to the place of energy crop cultivation. Energy crops for biogas production are cultivated only in Germany due to specific EEG feed-in tariffs of the past. For the year 2007, statistical data from Baden-Württemberg and Rhineland-Palatinate are available at the district level on the cropping areas for biogas production (SL BW 2014c; SL RLP 2009, p. 105). Since 2007, the number of agricultural biogas plants has strongly increased. Based on the development of the number and electrical capacity of biogas plants (DLR Eifel 2012, p. 15, MLRV BW 2014), the energy crop area for biogas was extrapolated for the year 2013. The year 2013 and not the baseline year 2010 is selected because the biogas plants in the year 2013 will be in operation more or less during the period under review in the project 'OUI Biomasse' due to the guaranteed feed-in-tariffs for 20 years. Additionally, it is assumed that the feedstock composition has not changed from 2007 to 2013 and that the efficiency of the biogas plants is constant.

A "virtual land use" was calculated for the energy crop production used for biofuels. "Virtual land use" represents the contribution of the URR – at national average – to the French respective German land use for biofuel production, independent from regional origin of the biofuel feedstuff in reality. In other words, this virtual land use intends to represent the part of rapeseed, sunflower, wheat, barley, rye, maize and sugar beet production area which is dedicated to biofuel production, irrespective of direct sales of these productions in the URR into biofuel production or of compensating higher shares for biofuel production in other regions. For Germany, the national energy crop areas are based on the areas published by FNR (2014a + b), normalising greater variations of the last years. For France, the national areas and biofuel production are derived from the heterogeneous data of ADEME (2012) and Agreste (2009). Due to missing data on French bioethanol crop areas, these areas are calculated from the feedstock input data of ADEME (2012, p. 30) and average yields for these crops (Agreste 2014). The overall percentages of national energy crop area are used to separate the energy crop areas in the German and French URR districts from the respective conventional arable cropping areas.

3.1.1.2 Land use change

For the scenario analysis (section 4.1), the agricultural land available in the year 2030 has to be estimated. The land use change in the period from 2000 to 2010 is used as a baseline for future land use change in the URR until 2030. The data for past agricultural land use change is taken from farm structure and/or land use surveys of the statistical offices (BFS 2013a, DRAAF Alsace 2014, Insee 2014, SL BW 2011a + 2014b, SL RLP 2005 + 2012a). For a better understanding of the drivers behind the agricultural land use change, the development of settlement and transport area was also investigated for the period 2000 – 2010 (Agreste 2010 + 2011, BFS 2013d, LEL 2014, SL RLP 2003 + 2011). Future reduction of UAA is restricted to the land use change triggered by settlement and transport area expansion. It is assumed that the slowdown trend of increasing settlement and transport area during the last decade (Destatis 2014, p. 15) will continue in the future and the sustainability goal of a further reduction of land consumption will be achieved. In consequence, the rate of UAA reduction in the period 2000 – 2010 is halved for the period 2010 – 2030. The development of UAA is assigned solely to reduction of arable land. Changes between permanent grassland and arable land are determined by agricultural policy regulations and scenario dependent.

3.1.1.3 Agricultural yields and production

The yield of arable crops for the reference year 2010 is determined as the average yield of the years 2008 – 2012. For the German part of the URR, crop yields at the level of district (Landkreis) are used (Statistische Ämter 2014). In the case of missing years, an average of the remaining years is calculated; for missing yield data for some crops in some districts, the yields of the neighbouring district are applied. For the French part, crop yields are only available at the level of département (Insee 2014c). These yields are used equally for all arrondissements. For crops not included in the data base, corresponding yields from German districts are adopted. For Switzerland, only yield data at the national level (SBV 2013) and at Kanton level for BS/BL and Aargau (SBV 2012) were found. In case of missing data at Kanton level, the average Swiss figures are used. All these yield data are regarded as yields of conventional farming. For organic farming, statistical data on crop yields in the URR do not exist. Based on different meta-analysis about organic to conventional yield ratios (Badgley et al. 2007, de Ponti et al. 2012, Seufert et al. 2012), it was assumed that organic yields for all crops are 80% of the conventional yields.

Future yields are extrapolated from the yield developments from 2000 to 2010, calculated as average yields of the years 1999 – 2003 and the years 2008 – 2012 for the German sub-region. For the two French départements, the average of the years 2000-2003 is used, data was provided by DRAAF Alsace (DRAAF 2014b). Yields in Switzerland remained since 2000 on the same level with annual fluctuations (BFS 2009, p. 20; BFS 2014). It is assumed that yields will remain unchanged until 2030 due to the specific Swiss arrangement of agricultural policies. For the reference year 2010 and for the year 2030 in the scenarios, the production is calculated from yields and arable crop areas, separately for both conventional and organic food and feed production, as well as energy crop production (on conventional arable land).

3.1.1.4 Residues from crop production

Available biomass residues are calculated for small-grain cereals (wheat, barley, rye, oat, other cereals), corn maize, oil crops (rapeseed, sunflower, soy) and sugar beet. The assessment of residues from crop production starts with crop specific residue to product ratios (respectively straw-grain ratios), taken from the German "Düngerverordnung" (2007). In the next step, fixed factors for extractable residues (Table 3.1) are applied, reflecting the amount of organic matter necessary to remain on the field for maintaining soil fertility.

Subsequently, the extractable small-grain straw is reduced by the straw uses in animal husbandry. Data of straw feeding amount are only available at the national level of Germany. A German average ration of 0,14 t straw per livestock unit for cattle, sheep and horses was calculated from the overall straw used as forage and the total number of livestock units (LSU) (BMELV 2012, p. 119, 130) and also used for French and Swiss URR districts. Small-grain straw for bedding is assessed from the livestock units of cattle, swine, sheep, goat and horse kept with solid manure systems and the specific straw bedding ratios. Data on the proportion of animal places in livestock husbandry systems with solid manure at the level of German Federal States (Baden-Württemberg, Rhineland-Palatinate) for cattle and swine (Destatis 2011) are used, and it is assumed that this ratio applies for all German URR administrative districts. Since no data was available for Switzerland and France, the average of the data for Baden-Württemberg and Rhineland-Palatinate was used for Switzerland and France. For sheep, goats and horses, 100% straw bedding for indoor animal husbandry is assumed. For sheep and goats, a free range period of 250 days, and for horses, a free range period of 105 days is applied (Gauder et al. 2011). The applied ratios of litter straw are based on the standard values of LfL 2013. The parameters for the dry matter content (or moisture content, respecttively) are also taken from the German Düngerverordnung (2007).

Table 3.1: Percentage of extractable crop residues.

Crops	Proportion of extractable residues (%)		
	Actual 2010Scenario BAU 2030	Scenario MaxEx 2030	Scenario ConsRec 2030
Small-grain cereals	50	60	40
Corn maize	40	40	30
Oil crops	50	60	40
Sugar beet	40	50	30

3.1.1.5 Manure

Manure potentials for biogas production include cattle, swine and poultry husbandry and are separately calculated for slurry and dung systems. The livestock numbers are taken from BFS (2014a), DRAAF (2012b), SL BW (2011c) and SL RLP (2012c). Average manure production per livestock unit is based on data from LTZ 2011, which includes the amount of liquid and solid manure derived for one livestock unit per day, data for the dry matter content of solid manure of cattle was found in FNR (2010, p.77) and was used for both cattle and swine, FNR also provided data for dry mass content of liquid manure. Manure production in farms with small number of livestock units (LSU) is excluded. For Germany, only farms with more than 50 LSU cattle or more than 100 LSU swine or poultry are included. These thresholds are based on Seyfert et al. (2011, p. 19). Livestock size classes are only available at the level of federal states (Destatis 2011) and the ratio of livestock over the threshold for Baden-Württemberg respectively Rhineland-Palatinate is applied equally for the respective German URR districts. For Switzerland, information on livestock size of farms is not accessible. Instead, 50% of the theoretical potential is regarded as available for an energetic utilisation, as assessed from Steubing et al. (2010). For the Alsace region, the German average percentage of livestock over the livestock size threshold was used, since no other data was available.

3.1.2 Approach to assess forestry land use and biomass production data

The assessment is based on forestry statistics for current forestry land use and wood harvest and on forest inventories for additional forest wood residues available for bioenergy production.

3.1.2.1 Forest areas

Forest area data for the reference year 2010 are taken from BFS (2011), DRAAF Alsace (2011), SL BW (2015) and SL RLP (2011). For the German and Swiss part of the URR, area data are available at district level; for the French part, only at the level of département (based on CORINE land cover classification 2006).

The slope of forest areas can restrict the possibility of harvesting wood. For the potential assessment, the analysis starts with 18 slope classes (Kappler 2008) which are summarised to five slope categories: "easy", "possible", "hard", "extreme" and "impossible". Of these only the first three classes are regarded as harvestable without causing too much effort and costs. Consequently, the classes "extreme" and "impossible" are excluded from the scope of the potential analysis. The areas of the slope classes are derived from freely available digital terrain models with 90m resolution (SRTM). Up to 60% of the coniferous forests are not utilisable; of broadleaf forests the percentage only reaches up to 41%.

Additionally, forest areas are distinguished between public or private ownership. Wood harvest is higher in public forests as in private forests. The distinction between public or private forests is needed for the different mobilization rates of wood residues. The share of private forests in Alsace is taken from FIBOIS 2015 and is quite equal (overall 25%). For the German part of the URR, the share is calculated from the forest inventory (Thünen Institute 2014), 27% of the forests in Rhineland-Palatinate and 36% in in Baden-Württemberg privately owned. For Northwest Switzerland, the share comes from "Forschungsanstalt WSL" (2014), with 19-22% private forests.

Protected areas only make up for a small share of the total forest area in the Upper Rhine Region with a high uncertainty in the location of these areas. Additionally, the utilisation of protected areas is dependent from the concrete on site regulation of the forest reserve and partly possible. Consequently, the protected areas are excluded from the scope of this approach.

3.1.2.2 Current wood harvest

Wood harvest is classified in stem wood, industrial wood and energy wood. An average wood harvest for a five year period – the period 2009 – 2013 was used for Alsace and Northwest Switzerland and the period 2008 – 2012 for the URR parts of Baden-Württemberg and Rhineland-Palatinate, depending on the most current data available. For Baden-Württemberg and Northwest Switzerland, data are published at the level of Landkreis respectively Kanton (BFS 2010 – 2014, Forst BW 2008 – 2012). Only data for the two Département were found for Alsace (DRAAF Alsace 2015). Only data at the level of the federal state are available for Rhineland-Palatinate (SL RLP 2013). The wood harvest in the URR part Rhineland-Palatinate was calculated from the average wood harvest per hectare in Rhineland-Palatinate.

The increment of growth and extraction data from the forestry inventories are not directly comparable with wood harvest data from the forestry statistics (BAFU 2014, p. 31; Dieter, Englert 2005). The wood harvest data tend to underestimate the real wood harvest, but only the data from the forestry statistics gives information on the current energy wood production.

3.1.2.3 Wood residue potentials

Forest residues are distinguished into solid wood and non-solid wood such as leaves, needles etc. As the collection of the latter is more complex and costly than of solid wood, these amounts are excluded from this approach. The assessment of woody residue potentials is based on the annual increment from the national forest inventories (FIBOIS 2015, Forschungsanstalt WSL 2014, Thünen Institut 2014). The current annual increment is carried forward unchanged until 2030, and the percentages of coniferous and deciduous forest areas are held constant. Data are collected separately for coniferous and broadleaf forests. It is assumed that the total annual forest increment in the three lowest slope categories can be harvested. In the next step, the percentage (based on wet biomass in m^3) of solid forest residues from coniferous forests is assumed to be 8% and 20% from broadleaved forests according to Kappler (2008).

Subsequently, different mobilization rates for public and private forests are applied. According to Kappler (2008), a mobilization rate of 80% for public forests and of 60% for private forests is used. In a last step, the calculated residues are reduced by already used residues, with 65% for broadleaf wood residues and by 10% for coniferous wood residues, based on Kappler (2008). The result of these calculation steps (Figure 3.1) is wood residue potential in Volume (m^3). The density of a single wood type varies in a large range and as the composition of coniferous and broadleaf forests in private and public forests is quite complex. For the translation in dry matter, the density of the wood type with the largest share in the forests is used for simplicity reasons as an average for all wood types: spruce: 470 kg/m^3 for coniferous wood and beech: 720 kg/m^3 for deciduous wood.

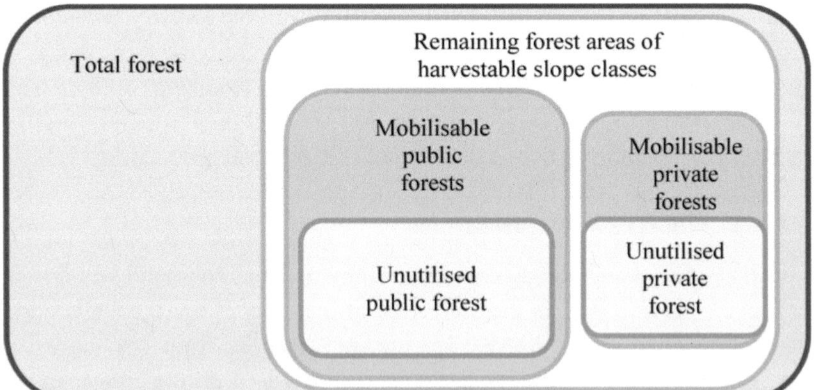

Figure 3.1: Methodology to estimate forest residue potentials.

3.1.2.4 Remote sensing data

Initially, the determination of forest biomass within this project had a strong relation to remote sensing data as a primary source of information. Given the data available for this task – basically, data provided by GISOR (IRS-P6) and USGS (Landsat), we must conclude that a reliable estimation is – if at all – only rudimentary possible. Within, this task we tested several approaches with the data we have: For a subset of Baden-Württemberg (Pine stands close by Karlsruhe), we tested machine learning approaches (random forest and support vector machines) with IRS-P6 data without success. For Alsace we applied a light use efficiency model to Landsat 8 data. However, the reference data basis seemed not to be strong enough to support sound conclusions of the computed results. A recent publication (Maak et al. 2016) indicate that for a large scale timber assessment, beside sophisticated remote sensing data (e.g. Lidar) and large scale reference data (NFI), additional predictors such as socio-economic factors are of importance. For future remote sensing approaches one must carefully evaluate the available data sets, to make sure that results with the desired precision can be achieved.

3.1.3 Approaches to assess waste biomass data

Waste biomass was estimated based on georeferenced statistical data. In this case waste statistics were used to estimate the total waste biomass potentially available. Waste statistics were provided either by the official statistical agencies in Germany, Switzerland, and the Conseil Général du Bas-Rhin, as well as the Syndicat Mixte Intercommunal de Traitement des Ordures Ménagères (SMITOM) in the department Haut-Rhin.

Waste statistics differ in all levels of the URR, administrative regions and types of waste. In some regions, statistics are given as total mass, in other regions values are in kg per Person. Data was transformed to absolute mass per administrative region and spatialised in masses per area. Data was collected for the year 2011. Statistical data was georeferenced with polygon shapes of the specific administrative regions. Spatial statistics values for cells of 500 x 500 m were calculated.

Household waste

"Household waste" is composite, non-recyclable waste derived from households. A review of household waste analyses from Germany and Austria in regions with separate collection of organic waste shows that household waste contains more than 30% of organic material (Elsaesser 2014).

Bulk waste

"Bulk waste" is a waste type which is too large to be accepted by the regular waste collection procedure. This waste is collected separately or delivered to the waste collection stations. With percentage fraction of >40% wood (UBA 2010), this type of waste is mainly useful for combustion.

Organic household waste

Organic household waste is defined as biodegradable municipal waste deriving from household waste. It is collected separately from other household wastes.

Green waste

In some regions, green waste from gardens is collected separately or delivered to the waste collection stations by the persons.

Vineyard residues

Three types of vineyard residues were assessed:

- Pomace is the solid remains of grapes after pressing for juice or oil. It contains the skins, pulp, seeds, and stems of the fruit.
- Yeast from fermentation of the wine.
- Prunings are the branches and leaves that are cut in the vineyards.

RLP Agroscience provided values for vineyard residues in t/ha for French vine cultivation areas. These values were spatialised with vine growing areas from Corine Landcover data. With the available sets of information discrimination between different vine growing regions and cultivars could not be done.

3.1.4 Challenges on data collection

In the previous subsections, a description of the process chain from plain data to biomass potential is elaborated in detail for the different relevant biomass types covered. It is clearly stated that we were obliged to make compromises in terms of accuracy and precision. The key challenges faced within this project are summarised out in the following passage.

Multi-nationality

The unique feature of the project OUI Biomasse of investigating a tri-national region including one none EU country turned out being one of the main challenges. Even though, the three countries are direct neighbours, approaches, definitions and political frameworks differ substantially. This diversity makes it almost impossible, to compare raw figures released by official bodies or private associations directly. There are always transformation, conversation and aggregation/dissolving steps involved which introduce errors and influence the final results.

Reference date & units

When comparing different sources, it is important that they match in terms of reference date and unit, e.g. some data is of unit tonnes per hectare whereas the other is in energy per hectare. Transformation involves the agreement on assumptions, such as wood density or energy content. The entire process introduces a considerable uncertainty, which is hard to overcome because essential information is usually missing. Furthermore it is very unlikely, that all collected data refers to the same date. Working with three countries most likely will result in three different reference dates. Aligning different temporal reference dates implicates interpolation or extrapolation, which also introduces ambiguity.

Data availabilities & policies

Biomass related data is collected by authorities and private institutions on regular basis. Annual enquiries are made in all domains such as national and regional forest inventories, waste management and by agricultural authorities. In fact, all relevant data which is necessary to make sound estimations on the development of biomass in the URR is actually collected and stored in different locations and institutions. However, it turned out that these data is almost not accessible for our purposes. Either a general legal restriction applies, so the data cannot be given away (because of e.g. privacy concerns) or data holders charge for the (public) data or restrict the usage by contract. Even institutions which purpose is the coherent collection of GIS-data in the TMO had us to sign several contracts and finally delivered only a fraction of the data they hold. Here, all actors need to find a common understanding on what data is needed and make the benefits of sharing it for all involved parties transparent and attractive.

3.1.5 References section 3.1

ADEME (Agence de l'Environnement et de la Maîtrise de l'Energie) 2012, *Analyse rétrospective des interactions du développement des biocarburants en France avec l'évolution des marches français et mondiaux (productions agricoles, produits transformes et coproduits) et les changements d'affectation des sols*, Available from: <http://www.ademe.fr/sites/default/files/assets/documents/83115_rapport_final_fev_2012.pdf> [08.07.2015].

Agreste (Ministère de l'agriculture, de l'agroalimentaire et de la forêt) 2009, *Statistique agricole annuelle Résultats 2008 définitifs et 2009 semi-définitifs France métropolitaine*, Available from: <http://agreste.agri culture.gouv.fr/IMG/file/francemetropolitaine212.pdf>. [08.07.2015].

Agreste 2010, *Occupation du territoire en 2010*, Available from: <http://www.agreste.agriculture.gouv.fr/IMG/pdf_teruti2011T2-2.pdf>.[08.07.2015].

Agreste 2011, *L'utilisation du territoire en 2010*. Agreste primeur 260, Available from: <http://agreste.agriculture.gouv.fr/IMG/pdf_primeur260 .pdf>. [08.07.2015].

Agreste 2014, *Surfaces, productions, rendements de céréales, oléagineux et protéagineux, résultats 2012 et 2013 semi-définitifs*, Available from: <http://agreste.agriculture.gouv.fr/IMG/pdf/saa2014T2bspca.pdf>. [04.11.2014].

Agreste Alsace 2011, *Mémento de la statistique agricole*, Available from: <http://www.google.de/url?sa=t&rct=j&q=&esrc=s&source=web&cd=5&ved=0CEMQFjAE&url=http%3A%2F%2Fwww.alsaeco.com%2Findex.php%3Flang%3Dfr%26cmpref%3D38507666%26module%3Dmedia%26action%3DDisplay&ei=bjVrVPH8GMLbO4PRgMgM&usg=AFQjCNERMl7yi8tl_3PjSh_mYhK52lfcCQ&bvm=bv.79908130,d.ZWU>. [08.07.2015].

Badgley, C, Moghtader, J, Quintero, E, Zakem, E, Chappell, MJ, Avilés-Vázquez, K, Samulon, A, Perfecto, I 2007, Organic Agriculture and the global food supply. *Renewable Agriculture and Food Systems*, vol. 22, pp. 86-108.

BAFU (Bundesanstalt für Umwelt) 2014, *Jahrbuch Wald und Holz 2014*, Available from: <http://www.bafu.admin.ch/publikationen/publikation/01790/index.html?lang=de>.

BFS (Bundesamt für Statistik) 2010 – 2014, *Forstwirtschaft der Schweiz*. Taschenstatistik 2010, 2011, 2012, 2013, 2014, Available from: <http://www.bfs.admin.ch/bfs/portal/de/index/dienstleistungen/publikationen_statistik/publikationskatalog.html?publicationID=4561>. [08.07.2015].

BFS (Bundesamt für Statistik) 2013a, *Wichtigste Flächenkategorien*, Available from: http://www.bfs.admin.ch/bfs/portal/de/index/themen/07/03/blank/data/01/02.html. [08.07.2015].

BFS 2013b, *Flächen der Obstkulturen in Aren 2013*, Available from: <http://www.bfs.admin.ch/bfs/portal/de/index/themen/07/03/blank/data/01/02.html>. [08.07.2015].

BFS 2013c: *Rebfläche und Sorten 2013*, Available from: <http://www.bfs.admin.ch/bfs/portal/de/index/themen/07/03/blank/data/01/02.html>. [08.07.2015].

BFS 2013d: *Arealstatistik Standard – Kantone und Grossregionen nach 4 Hauptbereichen*, Available from: <http://www.bfs.admin.ch/bfs/portal/de/index/themen/02/03/blank/data/01.html>. [08.07.2015].

BFS 2014a: STAT-TAB: Die interaktive Statistikdatenbank. *Tabelle Landwirtschaftliche Betriebe nach Jahr, Kanton und Zone - Landwirtschaftliche Nutzfläche (LN) und Nutztiere auf Klassifizierungsebene 2*, Available from: https://www.pxweb.bfs.admin.ch/Selection.aspx?px_language=de&px_db=px-x-0702000000_102&px_tableid=px-x-0702000000_102\px-x-0702000000_102.px&px_type=PX.Z [08.07.2015].

BFS 2014b, STAT-TAB: Die interaktive Statistikdatenbank. *Tabelle Landwirtschaftliche Betriebe nach Jahr, Kanton und Zone - Landwirtschaftliche Nutzfläche (LN) und Nutztiere auf Klassifizierungsebene 3*, Available from: <https://www.pxweb.bfs.admin.ch/Selection.aspx?px_language=de&px_db=px-x-0702000000_103&px_tableid=px-x-0702000000_103\px-x-0702000000_103.px&px_type=PX>. [08.07.2015].

BMELV (Bundesministerium für Ernährung, Landwirtschaft und Verbraucherschutz) (Hrsg.) 2012, Statistisches Jahrbuch über Ernährung, Landwirtschaft und Forsten der Bundesrepublik Deutschland 2012. Landwirtschaftsverlag, Münster-Hiltrup.

De Ponti, T, Rijk, B, van Ittersum, MK 2012, 'The crop yield gap between organic and conventional agriculture', *Agricultural Systems*, vol. 108, pp. 1-9.

Destatis (Federal Statistical Office of Germany) 2011, *Wirtschaftsdünger, Stallhaltung, Weidehaltung*. Landwirtschaftszählung / Agrarstrukturerhebung 2010. Fachserie 3, Heft 6. Wiesbaden.

Destatis 2014, *Sustainable Development in Germany*. Indicator Report 2014, Available from: < https://www.destatis.de/EN/Publications/Specialized/

EnvironmentalEconomicAccounting/Indicators2014.pdf?__blob=public cationFile>. [08.07.2015].

Dieter, M, Englert, H 2005, *Gegenüberstellung und forstpolitische Diskussion unterschiedlicher Holzeinschlagsschätzungen für die Bundesrepublik Deutschland.* Arbeitsbericht des Instituts für Ökonomie 2005/2. Bundesforschungsanstalt für Forst- und Holzwirtschaft Hamburg.

DLR Eifel (Dienstleistungszentrum Ländlicher Raum Eifel, Beratungszentrum Nachwachsende Rohstoffe) 2012, *Biogasanlagen in Rheinland-Pfalz 2012*, 3. Betriebserhebung Biogas. Bitburg, Available from: <http://www.dlr.rlp.de/Internet/global/themen.nsf/0/e92b0dd27329 dcaac1257a680036470c/$FILE/Biogasbroschuere_2012.pdf>. [08.07.2015].

DRAAF Alsace 2011, *Un territoire boisé et parsemé de sols bâtis*, Available from: <http://draaf.alsace.agriculture.gouv.fr/IMG/pdf/2_Occup Sol_01_Foret_et_Zones_Baties_p14_15_cle0239fe.pdf>. [08.07.2015].

DRAAF (Direction Régionale de l'Alimentation, de l'Agriculture et de la Forêt d'ALSACE) 2012a, *Fiche Arrondissement*, Available from: <http://draaf.alsace.agriculture.gouv.fr/Fiches_arrondissements>. [08.07.2015].

DRAAF 2012b, *Fiche Canton*, Available from: <http://draaf.alsace.agricul ture.gouv.fr/Fiches-par-canton>. [08.07.2015].

DRAAF Alsace (Direction Régionale de l'Alimentation, de l'Agriculture et de la Forêt d'ALSACE) 2014, *GraphAgri Régions Edition 2014* (Parution), Available from: <http://draaf.alsace.agriculture.gouv.fr/IMG/pdf/Graph Agri_Alsace_2014_cle434ac6.pdf>. [08.07.2015].

DRAAF 2014b, *Rendements pour different fruits des champs en Alsace*. Personal communication with DRAAF 21.11.2014.

DRAAF Alsace 2015, *Information statistique, Publications forêt*, Available from: <http://draaf.alsace.agriculture.gouv.fr/Publications-foret>. [08.07.2015].

Düngeverordnung 2007, *Verordnung über die Anwendung von Düngemitteln, Bodenhilfsstoffen, Kultursubstraten und Pflanzenhilfsmitteln nach den Grundsätzen der guten fachlichen Praxis beim Düngen* (Düngeverordnung - DüV) in der Fassung der Bekanntmachung vom 27. Februar 2007 (BGBl. I S. 221), die zuletzt durch Artikel 5 Absatz 36 des Gesetzes vom 24. Februar 2012 (BGBl. I S. 212) geändert worden ist, Available from: <http://www.gesetze-im-internet.de/d_v/index.html>. [08.07.2015].

Elsaesser, David 2014, *Internal report on Task 1.3: Estimation of waste biomass for the Upper Rhine Region*, Interreg project OUI Biomasse, Universität Koblenz-Landau, Landau.

FNR 2014, *Basisdaten Bioenergie, Deutschland 2014*, Fachagentur Nachwachsende Rohstoffe, Gülzow, p. 40.

FNR (Fachagentur Nachwachsende Rohstoffe) (Hrsg.) 2010, *Leitfaden Biogas*. Von der Gewinnung zur Nutzung, Available from: <https://www.dbfz.de/fileadmin/user_upload/Berichte_Projektdatenbank/pdf_208-leitfaden_biogas_2010.pdf>. [08.07.2015].

FNR 2014a, *Anbau nachwachsender Rohstoffe in Deutschland für die Jahre 2013/2014*, Available from: <http://mediathek.fnr.de/media/downloadable/files/samples/a/b/abb03_rgb.jpg>. [25.11.2014].

FNR 2014b, *Anbau Nachwachsender Rohstoffe in Deutschland nach Kulturarten*. Personal communication with FNR 05.12.2014.

ForstBW 2008 – 2012, *Jahresbericht 2008, 2009, 2010, 2011, 2012 des Landesbetriebes ForstBW*. 54., 55., 56., 57., 58. Jahrgang.

Gauder, M, Graeff-Hönninger, S, Claupein, W (2011): Identifying the regional straw potential for energetic use on the basis of statistical information. *Biomass and Bioenergy*, vol. 35, pp. 1646-1654.

Insee (Institut national de la statistique et des études économique) 2014a, *Utilisation du sol*, Available from: <http://www.insee.fr/fr/themes/tableau.asp?reg_id=15&ref_id=agrtc10201>. [08.07.2015].

Insee 2014b, *Production agriculture biologique*, Available from: <http://www.insee.fr/fr/themes/tableau.asp?reg_id=15&ref_id=agrtc10106>. [08.07.2015].

Insee 2014c, *Productions végétales principales en 2013* , Available from: <http://www.insee.fr/fr/themes/tableau.asp?reg_id=15&ref_id=agrtc10104>. [08.07.2015].

Kappler, GO 2008, *Systemanalytische Untersuchung zum Aufkommen und zur Bereitstellung von energetisch nutzbarem Reststroh und Waldrestholz in Baden-Württemberg*, Eine auf das Karlsruher bioliq®-Konzept ausgerichtete Standortanalyse.

Knop, W. 1868, *Der Kreislauf des Stoffs*, Lehrbuch der Agrikultur-Chemie. Haessel, Leipzig, p. 894.

Körner, I, Thrän, D, Scholwin, F 2009, *Exkremente aus der Nutztierhaltung*, Energie aus Biomasse, Grundlagen, Techniken und Verfahren, 2. Auflage, Hrsg.: Martin Kaltschmitt, Hans Hartmann, Hermann Hofbauer, Springer Verlag, p. 158.

LfL (Bayerische Landesanstalt für Landwirtschaft) 2013, *Rottemist- und Jaucheanfall verschiedener Tierarten in t bzw.* m^3 *pro mittlerem Jahresbestand in Abhängigkeit von Leistung, Fütterung und Haltungsform* (Stand. 2013, entspricht Düngeverordnung Anlage 5), Available from: <http://www.lfl.bayern.de/mam/cms07/iab/dateien/basisdaten_2013_6c.pdf>. [08.07.2015].

LTZ (Landwirtschaftliches Technologiezentrum Augustenberg) 2011, *Beratungsgrundlagen für die Düngung im Ackerbau und auf Grünland*, Available from: <http://www.ltz-bw.de/pb/site/lel/get/documents/MLR.LEL/PB5Documents/ltz_ka/Arbeitsfelder/Pflanzenbau/D%C3%BCngung/Versuche%20und%20Empfehlungen/BeratungsgrundlageD%C3%BCngung_DL/Beratungsgrundlagen%20f%C3%BCr%20die%20D%C3%BCngung%20im%20Ackerbau%20und%20auf%20Gr%C3%BCnland.pdf>. [08.07.2015].

MLRV BW (Ministerium für Ländlichen Raum und Verbraucherschutz Baden-Württemberg) 2014, *Entwicklung der Biogasanlagen in Baden-Württemberg sowie den Stadt-/Landkreisen*, Available from: <https://www.landwirtschaft-bw.info/pb/MLR.Landwirtschaft,Lde/Start seite/Erneuerbare+Energien/Biogas>. [08.07.2015].

MWVLW RP (Ministerium für Wirtschaft, Verkehr, Landwirtschaft und Weinbau Rheinland-Pfalz) 2008, *Ökologisch erzeugte Produkte*. Entwicklungsperspektiven für Erzeugung und Vermarktung in Rheinland-Pfalz, Available from: < http://mulewf.rlp.de/landwirtschaft/oekologischer-landbau/>. [08.07.2015].

LEL (Landesanstalt für Entwicklung der Landwirtschaft und der ländlichen Räume Schwäbisch Gmünd) 2011, *Strukturdaten zum ökologischen Landbau in Baden-Württemberg*, Available from: <https://www.landwirtschaft-bw.info/pb/MLR.LEL-ALT,Lde/Startseite/Nachhaltige+Unternehmens entwicklung/Strukturdaten+Oekolandbau>. [08.07.2015].

LEL 2015, *Bodenfläche nach Art der tatsächlichen Nutzung in Baden-Württemberg – Kreise – 1988 bis 2013*, Available from: <http://www.lel-bw.de/pb/,Lde/Startseite/Unsere+Themen/Strukturdaten_BW>. [08.07.2015].

OPABA (Organisation Professionnelle de l'Agriculture Biologique en Alsace) (2010): *L'observatoire de la production biologique alsacienne*. Les chiffres clés de 2010, Available from: <http://www.opaba.org/bioenalsace /wp-content/uploads/2013/01/LObservatoire-de-la-production-biologique-alsacienne-les-chiffres-de-2010.pdf>. [08.07.2015].

OPABA (Organisation Professionnelle de l'Agriculture Biologique en Alsace) 2013, *L'observatoire de la production biologique alsacienne*. Les chiffres de 2012, Available from: <http://www.opaba.org/bioenalsace/wp-content/uploads/2013/10/Opaba-24-pages-2013-bd.pdf> and <http://www.opaba.org/bioenalsace/agriculture-bio-en-alsace/les-chiffres-de-la-bio-en-alsace>. [08.07.2015].

SBV (Schweizer Bauernverband) 2012, 2.14 *Flächenerträge in ausgewählten Kantonen.*

SBV (Schweizer Bauernverband) 2013, 2.16 *Erträge der Pflanzenkulturen, ohne Obstbau*, Available from: <http://www.sbv-usp.ch/fileadmin/sbvuspch/06_Statistik/Pflanzen bau/se_2013_0216.pdf>. [08.07.2015].

Seufert, V, Ramankutty, N, Folex, JA 2012, 'Comparing the yields of organic and conventional agriculture', *Nature*, vol. 485, pp 229-233.

Seyfert, U, Bunzel, K, Thrän, D, Mauky, E, Fritsche, B, Schreiber, A, Liebetrau, J, Schmidt, T, Ulbricht, T, Lenz, V 2011, *DBFZ Report Nr. 8,* Deutsches Biomasseforschungszentrum, Leipzig, Available from: <https://www.dbfz.de/fileadmin/user_upload/DBFZ_Reports/dbfz_report_ kompakt_web.pdf>. [08.07.2015].

SL BW (Statistisches Landesamt Baden-Württemberg) 2011a, *Bodennutzung in den Stadt- und Landkreisen Baden-Württembergs 2010*, Available from: <http://www.statistik.baden-wuerttemberg.de/Veroeffentl/ Statistische_Berichte/3332_10001.pdf>. [08.07.2015].

SL BW (2011b): *Bodennutzung in den Stadt- und Landkreisen Baden-Württembergs 2010*. Endgültige Ergebnisse der Landwirtschaftszählung, Available from: <http://www.statistik.baden-wuerttemberg.de/ Veroeffentl/Statistische_Berichte/3332_10001.pdf>. [08.07.2015].

SL BW 2011c, *Viehhalter und –bestände in den Stadt- und Landkreisen Baden Württembergs 2010*. Ergebnisse der Landwirtschaftszählung. Statistische Berichte Baden-Württemberg Artikel-Nr. 3415 10001, Available from: <https://www.statistik-bw.de/Veroeffentl/Statistische_Berichte/ 3415_10001.pdf>. [08.07.2015].

SL BW (Statistisches Landesamt Baden-Württemberg) 2014a, *Struktur- und Regionaldatenbank, Landwirtschaft, Betriebliche Struktur, Ökologischer Landbau*, Available from: < http://www.statistik.baden-wuerttemberg.de/SRDB/home.asp?H=Landwirtschaft&U=01&E=GE>. [08.07.2015].

SL BW 2014b, *Regionaldaten. Landwirtschaftlich genutzte Fläche nach Hauptnutzungsarten 1999 und 2010*, Available from: <https://www.statistik-bw.de/SRDB/home.asp?R=KR337&H=Land wirtschaft&U=02&T=05025032#>. [08.07.2015].

SL BW 2014c, *Landwirtschaftliche Betriebe mit Anbauflächen zur Biogas-erzeugung als Hauptnutzung in den Stadt- und Landkreisen Baden-Württembergs 2007*. Personal communication from Dr. Anette Hartmann, 27.05.2014.

SL BW 2015, *Flächennutzung in den Stadt- und Landkreisen Baden-Württembergs 2010 nach Nutzungsarten*, Available from: <http://www.statistik.baden-wuerttemberg.de/BevoelkGebiet/Landesdaten/geb_flaeche_kreise_0000.asp?y=2010>.

SL RLP (Statistisches Landesamt Rheinland-Pfalz) 2003, *Nutzung der Bodenfläche 2003*. Ergebnisse der Flächenerhebung (tatsächliche Nutzung) , Available from: <http://www.statistik.rlp.de/fileadmin/dokumente/berichte/A5013_200300_1j_K.pdf>. [08.07.2015].

SL RLP 2005, *Agrarstrukturerhebung 2003*, Available from: <http://www.statistik.rlp.de/fileadmin/dokumente/berichte/C4633_200301_4j_K.pdf>. [08.07.2015].

SL RLP 2009, *Die Landwirtschaft 2008 mit Vergleichszahlen seit 1949*. Statistische Bände Band 397.

SL RLP 2011, *Nutzung der Bodenfläche zum 31. Dezember 2010. Ergebnisse der Flächenerhebung (tatsächliche Nutzung)*, Available from: <http://www.statistik.rlp.de/fileadmin/dokumente/berichte/A5013_201000_1j_K.pdf>. [08.07.2015].

SL RLP 2012, *Im Jahr 2012 wurden 3,6 Millionen Festmeter Holz eingeschlagen*. Knapp ein Fünftel als Energieholz genutzt, Available from: <http://www.statistik.rlp.de/wirtschaft/landwirtschaft/einzelansicht/archive/2013/april/article/im-jahr-2012-wurden-36-millionen-festmeter-holz-eingeschlagen/>. [08.07.2015].

SL RLP 2012a, *Agrarstrukturerhebung – Landwirtschaftszählung 2010, Bodennutzung und Viehhaltung der landwirtschaftlichen Betriebe*, Available from: <http://www.statistik.rlp.de/fileadmin/dokumente/berichte/C4633_201001_4j_K.pdf>. [08.07.2015].

SL RLP 2012b, *Agrarstrukturerhebung – Landwirtschaftszählung 2010. Ausgewählte Ergebnisse für ökologisch wirtschaftende landwirtschaftliche Betriebe*. Kennziffer C IV – 3j/10, Available from: <http://www.statistik.rlp.de/fileadmin/dokumente/berichte/C4663_201001_3j_K.pdf >. [08.07.2015].

SL RLP 2012c, *Viehbestände landwirtschaftlicher Betriebe 2010*. Endgültiges Ergebnis der Landwirtschaftszählung am 1. März, Available from: <http://www.statistik.rlp.de/fileadmin/dokumente/berichte/C3033_201001_ur_K.pdf>. [08.07.2015].

Statistische Ämter des Bundes und der Länder 2014, Regionalstatistik. *Hektarerträge ausgewählter landwirtschaftlicher Feldfrüchte* – Jahressumme – regionale Tiefe: Kreise und krfr. Städte, Available from: <https://www.regionalstatistik.de/genesis/online/data;jsessionid=DC878010AD41CADB011AE7232168FD16?operation=previous&levelindex=3&levelid=1413377717887&levelid=1413377342210&step=2>. [08.07.2015].

Steubing, B, Zah, R, Waeger, P, Ludwig, C 2010, Bioenergy in Switzerland: Assessing the domestic sustainable biomass potential. *Renewable and Sustainable Energy Reviews*, vol. 14, pp. 2256–2265.

Thünen-Institut 2014, *Dritte Bundeswaldinventur - Ergebnisdatenbank*, Available from: <https://bwi.info>. [17.12.2014].

WSL 2014, *Schweizerisches Landesforstinventar*, Available from: <https://www.lfi.ch/>. [30.10.2014].

UBA 2010, *Nutzung der Potenziale des biogenen Anteils im Abfall zur Energieerzeugung*, UFOPLAN 3733303. Umweltbundesamt Deutschland. Dessau.

3.2 Technologies, plant types and cost estimation

Christian Bidart*, Russell McKenna, and Wolf Fichtner*

French-German Institute for Environmental Research (DFIU), Karlsruhe Institute of Technology (KIT), Karlsruhe, Germany.

The conversion of the potential energy stored in biomass demands an in-depth understanding of its physical and chemical characteristics as well as the mechanisms that govern its transformations. The generation of useful forms of energy such as heat, electricity or fuels involve processes that are associated with an extraordinary level of complexity. Their comprehension is the result of a long road of research and experimentation, and the development and reformulation of theories that help to explain these transformations. This knowledge has become the cornerstone for developing technologies, which are nowadays an integral part of daily life in most industrialised societies.

The study of the chemical reactions and the physicochemical processes involved in biomass conversion can be seen as of paramount importance. The new knowledge arising from alternative approaches can be used to elucidate basic principles of the development of novel processes or as the scientific basis for the improvement of operation for proven ones. As usual in modern science, the creation of this knowledge is carried out layer by layer and a massive amount of work contributes only marginally to a given subfield.

Research Area three (RA3) of the 'OUI Biomasse' project consisted of four working groups which dealt with biomass conversion from different perspectives. They were groups from the *Laboratoire Gestion des Risques et Environnement* (**GRE**) from University of Upper Alsace, the *Institut de Chimie et Procédés pour l'Energie, l'Environnement et la Santé* (**ICPEES**) from Strasbourg University, *Génétique Moléculaire, Génomique, Microbi-*

ologie (**GMGM**) from Strasbourg University, and finally the *Deutsch-Französisches Institut für Umweltforschung* (**DFIU**) from Karlsruhe Institute of Technology. The main subjects of research were combustion, anaerobic digestion, methanation, and finally the techno-economic characterisation of conversion technologies respectively. The research of these four teams focused exclusively on biomass as feedstock, placing great attention on fundamental issues of its chemical transformation.

As far as **GRE** is concerned, a large portion of its work concentrated principally on demonstrating the feasibility of using grape marc residues for combustion on low scale (40 kW). Grape marc, a by-product from the wine industry of high relevance in the URR, can become a promising source of energy in domestic boilers; however, the emission of particles from its combustion raises environmental issues that will be addressed. Using an alternative approach, **ICPEES** conducted research on the methanation of grape residues in order to more efficiently produce methane as an energy carrier. To do so, ICPEES developed a Ni-ceria based catalyst, which was characterised in terms of reducibility, particle size and specific surface area. The results are promising in that this catalyst was shown to be effectively used. **DFIU** working group's main activity was the characterisation of biomass conversion processes in terms of their economics. This was conducted with the aim of supplying the information for solving the location problem that Research Area Four (RA4) had to deal with. The outcomes from RA4 analysis simultaneously became the starting point for the assessment of the impact of the utilisation of biomass. As the options of converting biomass are numerous, methodologies were established to select technologies considering scale of operation, type of biomass to be processed, end-products and other criteria. Finally, **GMGM** researched into the microbiological composition of grape marc and its correlation with the production of biomethane in order to find an alternative use of grape marc as a source of energy.

Section 3.2 contains the contributions of each working group to a knowledge and understanding of the process that can facilitate a more sustainable use of biomass in the URR.

3.2.1 Catalysts for syngas methanation

Anne-Cécile Roger, *Audrey Waldvogel and Claire Courson*

Laboratoire des Matériaux, Surfaces et Procédés pour la Catalyse (LMSPC), Université de Strasbourg, Strasbourg, France.

The reaction of CO methanation was first investigated to purify H_2-rich streams for ammoniac synthesis. Since the 60's this reaction aimed to produce Synthetic Natural Gas from coal through various types of processes based on coal gasification (*Kopyscinski et al. 2010*). More recently the carbon source shifted from coal to biomass or CO_2 (carbon recycling), leading to important changes in the processes. The catalysts used for the methanation of CO/CO_2 mixtures or for pure CO_2 methanation are generally not specific despite the fact that different mechanisms CO/H_2 and CO_2/H_2 may occur, allowing the development of efficient catalysts for such applications.

For biomass-to-methane processes, two main technologies of methanation sections are developed to manage the high exothermicity of the methanation reaction. In the Lurgi's type processes, a series of fixed bed reactors with high recycle rate and intermediate coolers are used to limit the adiabatic temperature below 450 °C (*Kopyscinski et al. 2010)*. For TREMP's process types, the adiabatic temperature is let to reach 700 °C (lower recycle rates are used) and the overheated steam produced is valorised into electricity (*Kopyscinski et al. 2010, Jensen et al. 2011)*. In such processes, the methanation catalysts have to be resistant to high temperature, more specifically resistant to deactivation by sintering, enhanced at elevated temperature (*Rostrup-Nielsen 2007, Nguyen 2013*).

Industrial methanation catalysts are generally based on the couple Ni-alumina. Promoters are added to increase their resistance to sintering and deactivation (*Kopyscinski et al. 2010, Ocampo et al. 2011*).

In the frame of 'OUI Biomasse' project, the methanation was studied to be on line on grape residues pyrolysis (studied by GRE, University of Haute Alsace), to efficiently produce methane as energy carrier.

3.2.1.1 Development of new methanation catalysts

The methanation catalyst, which has been developed at ICPEES, is composed of 5 wt% Ni°/ceria-based oxide. It has been prepared and characterised by numerous techniques. The main results are presented in Table 3.2.

The reducibility of both Ni-phase and oxide support phase were calculated from thermoprogrammed reduction experiments from room temperature to 900 °C. The particle size of both phases was determined by Debye-Sherrer equation from X-Rays Diffraction analysis. The pore volume and the specific surface area were determined by performing nitrogen adsorption-desorption at -196 °C using the Brunauer-Emmet-Teller (BET) method.

Table 3.2: Characterization of Ni/ceria-based catalyst.

	Ni reducibility (%)	Ce reducibility (%)	Support particle size (nm)	NiO particle size (nm)	Pore volume ($cm^3\ g^{-1}$)	Specific surface area ($m^2\ g^{-1}$)
Catalyst	100	74	9.8	25.7	0.42	49

The catalyst presents a total reducibility of NiO, which is totally reduced into Ni° after 4h at 400 °C under H_2/N_2 mixture (reduction procedure used to activate the catalytic material before performing catalytic tests). Ce^{4+} cations of the oxide support is partially reduced (74 %) into Ce^{3+}.The particle size of the support is around 10 nm whereas that of NiO is larger (25 nm). The catalyst present a BET surface around 50 $m^2\ g^{-1}$ and a pore volume of 0.42 $cm^3\ g^{-1}$.

3.2.1.2 Catalytic tests

The methanation reactions were first studied under COx/H_2 stoichiometric ratios. The preliminary results are presented in Figure 3.2 for CO_2/H_2

(ratio 1:4), CO/H_2 (ratio 1:3) and $CO/CO_2/H_2$ (ratio 0.6:0.4:3.4). The catalyst was initially reduced as detailed above. The tests were performed at atmospheric pressure under a GHSV of 43 000 h^{-1}. The results are presented in terms of COx conversion and CH_4 selectivity versus reaction temperature, and compared to the corresponding values expected at thermodynamic equilibrium.

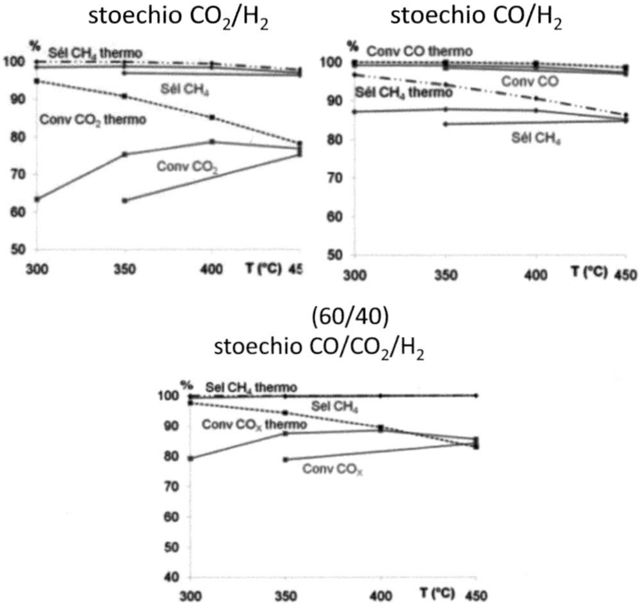

Figure 3.2: Methanation results under different stoichiometric conditions.

For CO_2 methanation, the methane selectivity is always close to 100 %, independently from the conversion. The CO_2 conversion is low at 300 °C (around 62 %), increases with temperature until 400°C (around 80 %) and finally slightly decreases at 450 °C where the thermodynamic limit is reached. After 16h at 450 °C under reaction conditions, a back point at 350 °C shows a deactivation.

For CO methanation the picture is rather different. Here the CO conversion is almost total over the whole test. The selectivity to methane is much lower than for CO_2 methanation, stable at a value around 80-85 %. Under CO/H_2, the undesired Water Gas Shift reaction is favoured at low temperature, leading to CO_2 as byproduct.

For CO/CO_2 co-methanation, an intermediate behaviour is observed accounting for the fact that no competition occurs between the two carbon oxides. Finally, comparable methane yields are obtained for the 3 conditions tested.

Conditions representative for syngas obtained by biomass pyrolysis, where highly H_2-deficient mixtures are produced, were first taken from litterature. The composition of syngas obtained from bagasse pyrolysis is reported in Table 3.3 at different pyrolisis temperature. The CO/CO_2 ratio, the H_2/COx ration as well as the H_2 content with respect to H_2 theoretical amount needed for methanation of the whole COx molecules are also reported.

Table 3.3: Syngas composition after bagasse pyrolysis.

T (°C)	H_2 (mol/ $kg_{bagasse}$) 10^3	CO (mol/ $kg_{bagasse}$) 10^3	CO_2 (mol/ $kg_{bagasse}$) 10^3	H_2/ COx	CO/ CO_2	H_2/ H_{2theo} (%)
400	1.6	502	2008	0.0006	0.25	0.017
500	32	567	2537	0.01	0.22	0.27
600	225	636	2222	0.079	0.29	2
700	1272	774	1840	0.48	0.42	13
800	2475	1165	1615	0.89	0.72	25
900	3118	1385	999	1.31	1.39	38

The flow of syngas produced by bagasse pyrolysis increases with temperature. Despite the fact that H_2 production becomes important with temperature, it clearly appears that even at high temperature a strong hydrogen deficiency towards methanation is obtained. The best hydrogen content is reached at 900 °C but only corresponds to 38 % of what would be needed for total methanation of CO/CO_2 flow.

The pyrolysis of wine residues at 800 °C studied at GRE in Mulhouse gave the following results : the H_2/COx ratio was equal to 0.78 and the hydrogen content with respect to stoichiometry was 20.4 %. At 900 °C, the H_2/COx ratio was equal to 1.03 and the hydrogen content with respect to stoichiometry reached 28.4 %. That shows that similar results are obtained comparing bagasse and wine residues pyrolysis process.

Based on these different results, methanation conditions were chosen to simulate the on-line syngas methanation. The different catalytic tests conditions are listed in Table 3.4. Note that a stable nitrogen flow was added in the gas mixture to serve as internal standard for the carbon balance.

Table 3.4: Experimental conditions of syngas methanation.

Test	GHSV (h^{-1})	$m_{catalyst}$ (g)	N_2 (mL min^{-1})	He (mL min^{-1})	H_2 (mL min^{-1})	CO_2 (mL min^{-1})	CO (mL min^{-1})
1	43 000	72.9	10	/	36	9	/
2	43 000	43.1	10	/	13.5	9	/
3	43 000	43.1	10	11.25	6.75	4.5	/
4	43 000	81.2	10	40	6.75	4.5	/
5	81 200	43.1	10	40	6.75	4.5	/
6	81 200	45.7	10	35.6	10.2	5.7	3.3

The first test corresponds to the test of CO_2 methanation presented in Figure 3.2. Test 2 will simulate the hydrogen deficiency which is to be expected from wine residue pyrolysis (H_2/CO_2 ratio was decreased from 4 to 1.5). Note that the active phase of methanation being metallic Ni°, an excess of CO_2 could reoxidise the catalyst and lead to deactivation. For tests 3 and 4, the aim is to study the effect of dilution. In Test 5, the value of GHSV is almost twice as for test 3 (performed in the same component ratios). For test 9 the compositions obtained from wine residues pyrolysis were simulated.

The catalytic results of tests 1 to 5 are presented in Figure 3.3 (H_2 conversion, limiting reactant) and Figure 3.4 (CH_4 selectivity).

For all the tests the H_2 conversion rapidly increases with temperature between 200 and 300 °C. The hydrogen defiency does not affect much the catalyst reactivity in terms of concersion and selectivity. The catalyst is able to stand both high hydrogen deficiency and high gas space velocity. A slight decrease of conversion and selectivity at very high GHSV can be noted (Test 5). The catalyst does not seem to be deactivated by CO_2 excess.

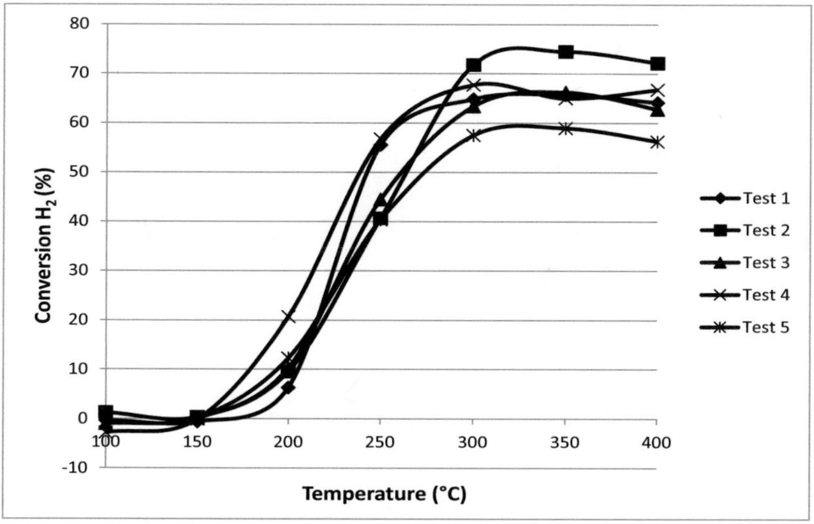

Figure 3.3: Hydrogen conversion obtained for tests 1 to 5.

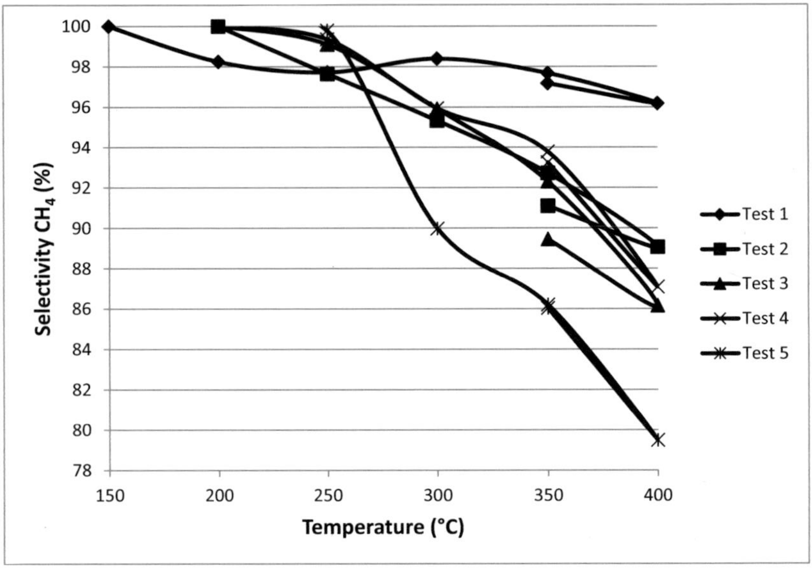

Figure 3.4: Methane selectivity obtained for tests 1 to 5.

In summary, the results obtained for tests 2, 3 and 4 are very comparable. The catalytic behaviour is thus poorly affected by the dilution. The selectivity to methane is however lower than that obtained for Test 1.

At higher GHSV (Test 5) the effect is even more pronounced.

The results of test 6 which simulates the syngas composition expected from wine residues pyrolysis at high temperature are presented in Figure 3.5 and Figure 3.6. Here again good H_2 conversion and methane selectivity are obtained. The best methane yields can be obtained at a reaction temperature around 350 °C.

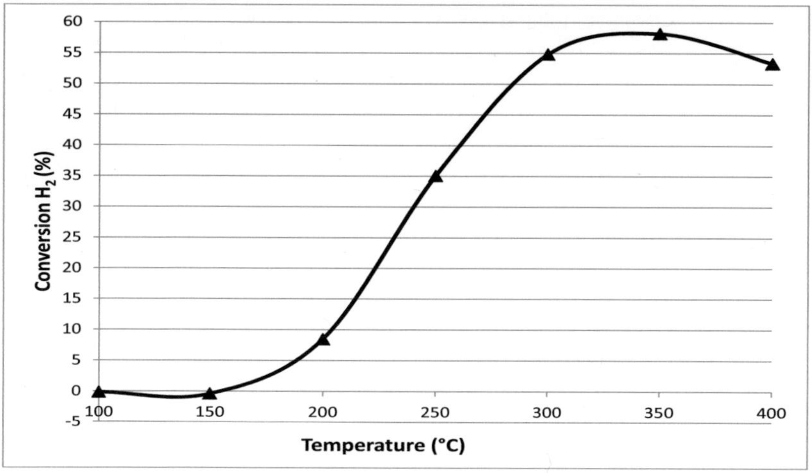

Figure 3.5: Hydrogen conversion obtained for test 6.

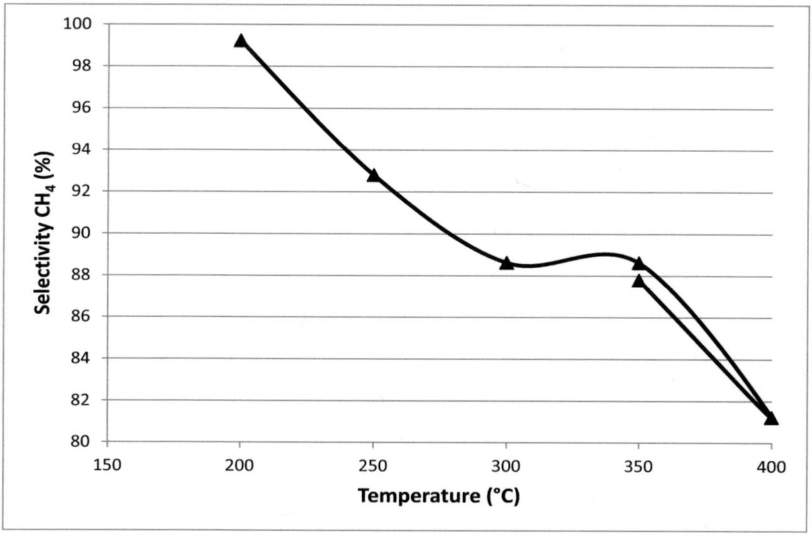

Figure 3.6: Methane selectivity obtained for test 6.

Preliminary results indicated that the Ni/ceria-based catalyst developed at ICPEES is active for co-methanation. The last results confirm that, even at high velocity and high hydrogen deficiency, and under conditions simulating the syngas composition expected from wine residues pyrolysis at high temperature, the developed catalyst is efficient and stable. This catalytic material is then totally adapted to be used in a methanation reaction, online with a pyrolysis unit for the energetic valorisation of wine residues.

3.2.2 Investigation of technical usability of wine residues

Maximilien Valente, *Cornelius Schönnenbeck, Gwenaëlle Trouvé, Alain Brillard, Jean-Francois Brilhac*

Laboratoire de Gestion des Risques et Environnement (GRE), Université de Haute Alsace, Mulhouse, France.

The Alsace Region (France) is known for wine production (15 600 hectares) leading to large quantities of grape marc. Using the factors given by Toscano *et al.* (2013), the calculated corresponding dry matter is 12 000 tons per years. Different pathways (Saidur *et al.* 2011) are developed for the valorization of the wine residues (distillery, large scale thermal facility for energy recovery, *etc.*). Beside them, wine residue presents a great interest for energy recovery. Celma *et al.* (2007) assessed the use of grape marc in Extremadura (Spain) and calculated good economical specific costs and on site availabilities for this surrogate fuel. Toscano *et al.* (2013) examined the physical-chemical composition (especially carbon content) of grape marc residues in Italy and exhibited values close to wood chips, beech chips or wheat straw. Considering biomass combustion, a lot of recent works point out domestic boilers as an important source of particles emissions. Favez *et al.* (2009) demonstrated that in Paris, during winter, 20±10% of the ambient $PM_{2.5}$ mass can be attributed to biomass burning.

Hence a constant assessment of combustion quality, especially at the domestic boiler scale, has to be performed in order to limit pollutants emission.

Burning of wine residues in a domestic biomass boiler can be considered as an alternative pathway for grape marc recovery. The winemakers could then use directly part of the waste produced during wine production as a fuel. The present contribution is devoted to the study of grape marc combustion for a low environmental impact. To our knowledge, this study is the first considering grape marc combustion in a domestic scale boiler. The aim is to characterise the thermal degradation of grape marc by thermogravimetric analysis and to analyse the energy recovery in a domestic boiler focusing on the emission factors of gas and particulate matter.

3.2.2.1 Material and methods

Samples and characterisation

Wine residues were collected directly after the pressing step and delivered by a winemaker located in Orshwir (Alsace, France). For the thermogravimetric analysis (TGA), three different varieties of grape marc were tested: Sylvaner, Chasselat and Pinot Gris. Gewürztraminer variety was used for the experiments in the multi-fuel boiler. In addition, two different biomasses were used in this study for multi-fuels experiments in the boiler: miscanthus giganteus locally produced at Ammertzwiller (Alsace, France) and DIN CERTCO pellets, purchased from SOFAG (Arc sous Cicon, France). Physical and chemical characterisations of the different samples were performed. Ultimate analysis was carried out to determine carbon, hydrogen, oxygen, nitrogen and sulfur weight fractions. Table 3.5 gives the results for Gewürztraminer variety. The results obtained for the other grape marc varieties are presented in Valente *et al.* (2015).

Samples were characterised following the standard XP CEN/TS for the moisture, ash contents and the low heating value (LHV). A calorimetric bomb (IKA) was used to determine the high heating value (HHV). Low heating values (LHV) were calculated based on free ash and water contents.

Table 3.5: Fuel characteristics.

	LHV $MJ.kg^{-1}$	Moisture %	Ash[c] %	Ultimate Analysis on dry basis (%)				
				C	H	O	N	S
Pellets (DIN CERTCO)	17.7	8	0.3	47.1	6.1	46.6[b]	<0.1	<0.1
Pellets/Grape 66/33	13.2[a]	27	0.8	48.0[a]	6.3[a]	n.d.	n.d.	n.d.
Pellets/Grape 50/50	10.9[a]	37.5[a]	n.d.	48.6[a]	6.3[a]	n.d.	n.d.	n.d.
Grape (27% humidity)	14	27	0.8	50.04	6.42	36.48[b]	2.49	<0.3
Beech wood chips	16.5	5	0.4	47.6[d]	5.7[d]	44.8[d]	0.42[d]	0.01[d]
Miscanthus	16.7	10	1.4	47.7	5.89	43.7	<0.3	0.12
Misc/Grape 75/25	13.5[a]	24.25[a]	n.d.	49[a]	6[a]	n.d.	n.d.	n.d.
Misc/Grape 66/33	12.5[a]	32	1.5	49.1[a]	6.1[a]	n.d.	n.d.	n.d.
Misc/Grape 50/50	10.4[a]	38.5[a]	n.d.	49.4[a]	6.2[a]	n.d.	n.d.	n.d.
Grape (67 % humidity)	4	67	0.4[a]	50.04	6.42	36.48[b]	2.49	<0.3

[a]: calculated, [b]: calculated by difference, [c]: on raw basis, [d]: www.ecn.nl/phyllis2/Browse/Standard/ECN-Phyllis Beech #64, n.d.: non determine

The boiler is devoted to multi-fuels presenting humidity up to 30-35%. Hence combustion tests of raw grape marc exhibiting a moisture content of 67% were not performed. Raw grape marc was then dried in order to reduce its humidity from 67% to 27%. In the current study, this dry grape marc was combusted in the REKA boiler and thermal and environmental performances were compared to these of DIN CERTCO pellets and miscanthus in preliminary experiments. Blends of miscanthus/dry grape marc and DIN pellets/dry grape marc in weight proportions 75/25, 66/33 and 50/50 were also prepared and tested in the REKA boiler. The fuel properties of the blends are given in the Table 3.5.

Thermogravimetric analysis (TGA)

The pyrolysis and the combustion of different parts of grape marc (skin, seed and stalk) were investigated by TGA in order to propose a relevant kinetic model and the corresponding kinetic parameters. For the last purpose, an extended Independent Parallel Reaction (IPR) model is proposed assuming that the biomass structure is decomposed according to the lignocellulosic material representation. The IPR kinetic model is indeed extended in order to take into account both the devolatilisation and the char combustion stages. Experiments were performed in a thermobalance TA Q500 at heating rates of 5, 10, 20 and 50°C/min, in the temperature range 30–800°C, and under a synthetic air flow of 3.6 NL/h (20% O2, 80% N2). Some experiments were also performed under pure nitrogen. Small sample masses (10–30 mg) were used in order to reduce as far as possible mass and heat transfer limitations by diffusion inside the solid fuel bed.

Combustion tests in a multi-fuel boiler

Combustion tests were performed in a multi-fuel boiler (HKRST-FSK) supplied by REKA (Aars, Denmark) equipped for combustion studies. Performance of the boiler ranges from 30 to 40 kW. The boiler basis is a moving stepped grate. The step grate permits to use almost any biomass fuel since the grate movement prevents the possible slag formation. As the boiler and the fitted hopper were located on a balance, the mass loss and then the consumption of the biomass (pellets/ wood chips/ grape marc) were recorded. First experiments were performed using grape marc with 27% of relative humidity. Then combustion was performed with grape marc blended with wood pellets and miscanthus. Due to the heterogeneity of the samples it was not possible to stabilise the combustion of these blended samples. As a consequence new blended samples were prepared by grinding the grape marc with Miscanthus or Pellets and thus homogenizing the blended fuel. Thermal behaviour of the boiler was tested with samples at steady state during 30 min for experiments repeated at least three times. Considering the LHV and the heat losses during combustion, it is possible to evaluate the input and output powers. The combustion efficiency was then estimated using the NF EN 303-5 standard (revised 2012) related to boilers with a heating power lower than 500 kW and corresponding to the class 5. For fuels or blended fuels with low water content (up to 33%) the boiler efficiency was found always higher than 90% and this value is totally relevant considering the NF EN 12809 standard requiring an efficiency of at least 85%.

Gas emissions and particle matter analysis

Emissions were measured in the chimney according to EN-304 standard (NF EN 304/A1, 1998) during experiments on the REKA boiler. Flue gas temperatures were continuously recorded. O_2, CO, CO_2 and NO_x were analysed respectively by paramagnetic and specific infrared analysers (ROSEMOUNT NGA 2000). According to EN 303-5 (NF EN 303-5, 2012) standard, concentrations expressed in mg. Nm^{-3} were referred to 10% of O_2 in the exhaust. An Electrical Low Pressure Impactor (ELPI) manufactured by DEKATI Ltd. (Tampere, Finland) was used to collect particles from 7 nm to 10 μm into 12 size fractions. Particle number and concentrations calculated accounts for this particularity using 1 g. cm^{-3} of density value (Marjamäki et al., 2000). The total mass fraction of Total Suspended Particles (TSP) was measured by gravimetry according to DIN CERTCO certification rules (DIN CERTCO, 2008).

3.2.2.2 Results and discussion

Investigation of thermal degradation of grape marc by TGA

Experiments were performed with each part of the grape marc (skin, seed and stalk) under air and pure nitrogen from ambient temperature to 950°C. Figure 3.7 reports TGA and DTG curves for the different parts of the grape marc for Sylvaner variety. The DTG analysis shows that each constituent reacts at different temperatures. The devolatilisation steps occur from 150°C to 400°C. The different peaks observed correspond to the degradation of Hemicellulose, Cellulose and Lignin, respectively. The char combustion step (from 400°C to 600°C) starts earlier and is faster for stalks (0.1 %/s at 401°C) than for the other parts. For grape skins and seeds, the oxidation peaks are about 0.06 %/s at 460 and 500°C respectively.

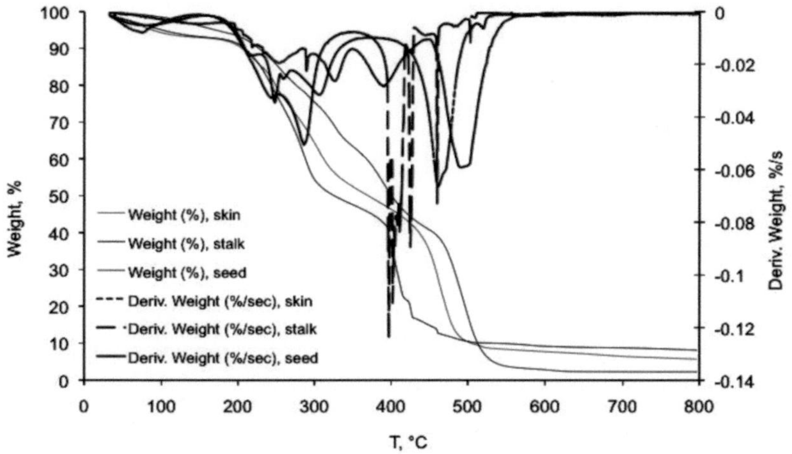

Figure 3.7: TG (Weight, %) and DTG (Deriv. Weight, %/s) for the different parts of Sylvaner grape marc (skin, seed and stalk); Heating rate: 5°C/mn under air. (Valente et al. 2015).

An Extended Independent Parallel Reaction (EIPR) model was proposed for the thermal decomposition of grape marc using its lignocellulosic representation under an oxidative atmosphere. The overall mass conversion rate during thermal degradation is expressed according to a first-order reaction system. This deterministic model compares the overall mass loss rates obtained from the experiments and from the model. The assumption and the equations of the model are presented in detail in Valente *et al.* (2015) and Brillard *et al.* (2015).

$$\frac{dm}{dt}(t) = \sum_{i=H,C,L} \left(\begin{array}{l} k_i(T(t)) \left(m_i(0) - \frac{1}{\tau_{vol,i}} m^e_{vol,i}(t) \right) \\ + k_{comb}(T(t)) \left(\frac{\tau_{char,i}}{\tau_{vol,i}} m^e_{vol,i}(t) - m^c_{char,i}(t) \right) P_{O_2} \end{array} \right).$$

A coupled system of six ordinary differential equations (ODE) is defined and has to be solved. The overall mass rate is then computed as: $m_i(0)$ is the mass of the volatiles and char contained in the constituent i at t=0, $m_{vol,i}{}^e(t)$ is the mass of volatiles emitted by the constituent i of the sample at time t, $m_{char,i}{}^c(t)$ represents the mass of char consumed at t among that produced from the constituent i of the sample (i=H, C, L). $\tau_{char,i}$ and $\tau_{vol,i}$ are respectively the volatiles and char mass fractions in constituent i. The kinetic parameters $k_i(T)$ and $k_{comb}(T)$ are supposed to obey Arrhenius equations:

$$k_i(T) = A_i \exp\left(-\frac{Eai}{RT}\right) \text{ and } k_{comb}(T) = A_{comb} \exp\left(-\frac{Ea_{comb}}{RT}\right)$$

The numerical resolution of the system of ODEs was done using the routine ode of the SCILAB software and the best kinetic constants were determined using the routine datafit. Table 3.6 presents the results obtained for the different parts of the grape marc at the heating rate of 5°C/min. The activation energy obtained for char combustion is consistent with literature data (see Valente *et al.* 2015) and is higher to these obtained for the thermal degradation of cellulose, hemicellulose and lignin.

Table 3.6: Kinetic parameters extracted for grape skins, stalks and seeds for Sylvaner variety (Brillard *et al.* 2015).

Skins	$m_i(0)$ (mg)	$\tau_{vol,i}$	A (s^{-1})	Ea (J/mol)
Hemicellulose	0.5000	0.6	25 000	70 000
Cellulose	2.0000	0.7	69 000	83 000
Lignin	2.2695	0.6	39	63 000
Combustion			6 000 000	190 000

Stalks	$m_i(0)$ (mg)	$\tau_{vol,i}$	A (s^{-1})	Ea (J/mol)
Hemicellulose	0.4500	0.7	25 000	73 000
Cellulose	3.6000	0.9	40 000	75 000
Lignin	7.2094	0.3	40	65 000
Combustion			5 000 000	172 000

Seeds	$m_i(0)$ (mg)	$\tau_{vol,i}$	A (s^{-1})	Ea (J/mol)
Hemicellulose	0.5000	0.6	25 000	70 000
Cellulose	3.5000	0.8	69 000	85 000
Lignin	16.0432	0.5	39	61 000
Combustion			1 000 000	190 000

As shown in Figure 3.8 for grape skin, a good agreement between calculated and experimental data is obtain with this set of kinetic parameter.

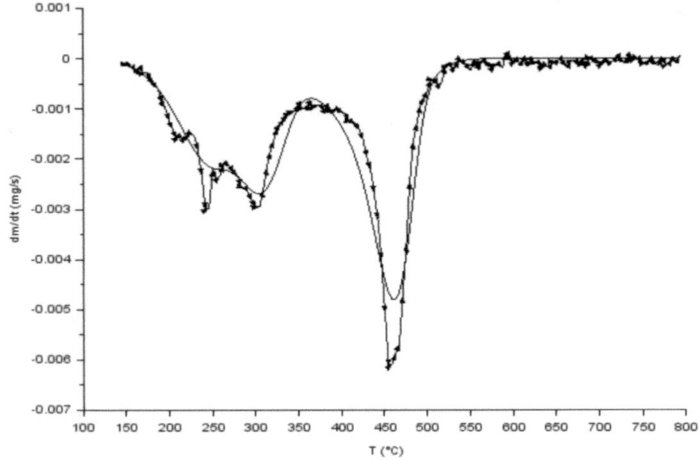

Figure 3.8: Grape skin DTG (Deriv. Weight, mg/s); 5°C/min. Experimental (symbol) and simulated (line) curves (Brillard et al. 2015).

3.2.2.3 Pollutants emissions for the combustion tests in a multi-fuel-boiler

Gaseous emissions and total suspended particles (TSP) concentrations

Exhaust emissions were recorded with the REKA multi-fuels boiler to estimate the impact of an addition of grape marc in the blended fuels. Figure 3.9 gives the average values of gas concentrations and TSP in the flue gas at steady state. The values obtained with conventional fuels such as miscanthus, beech chips and pellets are also presented for comparison. Limit values for the standard EN 303-5 are reported. CO concentrations are of the same order of magnitude than those described in literature for conventional dry fuels as pellets, miscanthus and beech chips (Verma *et al.*, 2011) and close to standard values. The combustion process is not significantly modified with an addition of grape marc since CO emission in the exhaust are still very low for blend ratio (%wt) miscanthus/ grape ranging from 75/25 to 67/33. For higher proportion of grape marc (50/50), CO concentration in the exhaust is higher and consequently the thermal performance of the boiler is affected due to the high relative humidity of the sample. Moreover, as shown in Figure 3.9, CO concentration is higher for the pellets/grape mixture than for the miscanthus/grape mixture whatever the mass ratio considered. This may be explain by how the grape marc is distributed in the biomass. Since miscanthus is constituted of fine particles, the grape marc is more homogeneously mixed in the miscanthus/grape mixture than in the pellets/grape mixture. The shape of the natural biomasses used acts strongly on combustion efficiency. The reactive surface in the miscanthus blends may be higher than for pellet blends. This enhances the combustion. Excepted for the DIN CERTCO pellets, mean values of NOx are higher than standards. This result may be linked to the relatively high nitrogen content in the fuels.

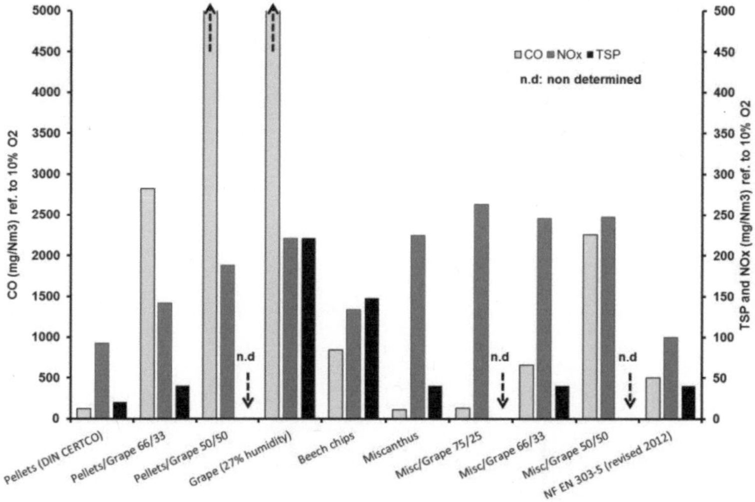

Figure 3.9: Concentrations of CO, NOx and TSP compared to standards.

The total mass concentration of TSP is presented at 10% O_2 in Figure 3.9. TSP is ranging from 20 to 40 mg Nm^{-3} for most of the fuel tested excepted for grape marc with 27% of humidity and beech chips with 220 and 150 mg Nm^{-3}, respectively. For conventional biomass such as pellets and miscanthus but also for blends with a proportion of marc equal to 33 % by weight, TSP are close to the limit values of the actual standard EN 303-5. One can conclude that the presence of 33% of grape marc in the blends has no significant influence on dust emission. However for higher content of grape marc in the blends, the emission of dust and CO will be surely higher.

Fine and ultrafine particle emissions

Figure 3.10 gives the total number PM_{10} size distributions for raw and blended miscanthus samples. Quite similar distributions were also obtained for raw and blended pellets samples. The aerosols emitted during the combustion process of these samples primarily consisted of fine submicron particles with a nucleation mode dominating. The number distribution for raw pellets and miscanthus are centered between 0.12 and 0.2 µm. These distributions are very close to data found in literature during combustion of various woody fuels and biomasses as energy crops (Dorge *et al.* 2011; Johansson *et al.* 2004).

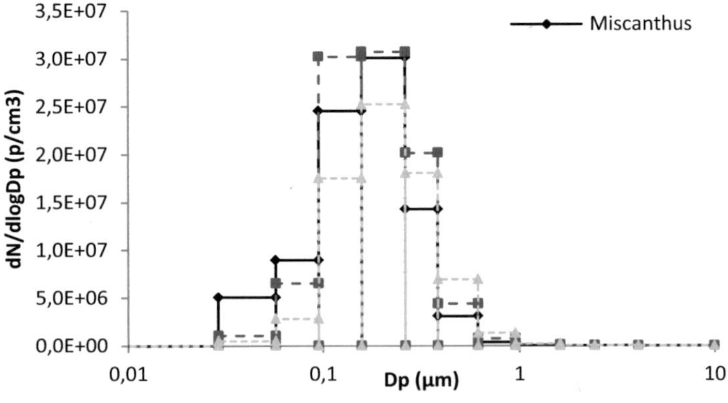

Figure 3.10: Influence of grape marc proportion to the number size distribution of particles emitted during the combustion of Miscanthus (Valente *et al.* 2014).

Addition of grape marc in miscanthus slightly modifies the number size distribution as described in Figure 3.10. This distribution is moved toward bigger particles. However, particles with diameters in the range 0.1 to 1 μm still dominate. The fraction $PM_{2.5-10}$ is totally negligible. Particle matters ($PM_{0.1}$ to $PM_{2.5}$) recorded during tests show that the main particles emitted are ultrafine particles ($PM_{0.1}$ and $PM_{0.1-1}$). PM_1 represents more than 99% of the total number of particles for all samples. The fraction $PM_{1-2.5}$ is below 0.1% of the total number of particle. From the number size distribution presented in Figure 3.10, the total $PM_{2.5}$ number concentration is calculated. It varies from 1.0 10^{13} particles per m^{-3} to 1.5 10^{13} particles per Nm^{-3} related to 10% of O_2 in the flue gas for raw pellets and miscanthus, respectively. Considering a relative standard deviation close to 20%, the total $PM_{2.5}$ number concentration is not significantly impacted by the addition of grape marc in the miscanthus blends.

3.2.2.4 Conclusion of subsection 3.2.2

The use of grape marc for energy recovery could be relevant for the wine makers as it is renewable and has an almost neutral CO_2 balance. This study is the first demonstrating the feasibility of grape marc combustion in a small domestic scale boiler (40kW).

The first part of the work was devoted to the investigation of the thermal decomposition of the grape marc by TGA. The main theoretical contribution and input was to characterise accurately this biomass (Valente *et al.* 2015) and to develop a model simulating its decomposition including the reaction of char combustion (Brillard *et al.* 2015). Corresponding kinetic parameters were proposed for each part of the grape marc.

The investigation of grape marc combustion in a multi-fuel boiler was then performed. Concerning pollutant emissions, this study pointed out the influence of the humidity of raw wine residues on the thermal process and consequently on the emission of gaseous pollutants and particulate matter. The moisture content of raw grape marc (between 65-80%) is an issue for the combustion process since it leads to high concentrations of carbon monoxide in the fumes as well as volatile organic compounds and particles in the exhaust line. Hence, only blended fuels Miscanthus/Grape marc and Pellets/Grape marc are relevant for combustion. Using such a blended fuels, the emission of gas and particulate matter are more or less similar to these observed for other biomass resources or co-products. Such a result indicates that the use of grape marc in a multi-fuel system is fully relevant for energy recovery.

Further studies of emissions factors of grape marc burning regarding air excesses, optimized fuel appearance, or perhaps a pelleting process could lead to an optimisation of the combustion process and therefore a new source of energy in the Alsace region but also throughout the world in vineyard areas.

3.2.3 Biomass conversion technologies

***Christian Bidart**, Russell McKenna and Wolf Fichtner*

French-German Institute for Environmental Research (DFIU), Karlsruhe Institute of Technology (KIT), Karlsruhe, Germany.

The three uses of biomass for energy utilisation are basically heating, fuels and electricity. These are useful forms of energy; they can, consequently, be distributed, traded and commercialised. Being produced on a large scale and traded at a relatively low price, they are normally categorised as commodities. The generation of any of these energy carriers demands generation processes which transform the primary energy (i.e. biomass, crude oil, natural gas, etc.) into secondary energy (i.e. heat, fuels or electricity).

Biomass can be classified into lignocellulosic biomass and biochemical biomass. Although these terms are not standard and differences can be found in definitions in the literature, they can be employed to denominate, for the former, biomass with a high lignin content, and for the latter, biomass that can undergo reaction by the action of enzymes, thus being converted by microorganisms. As Figure 3.11 shows, lignocellulosic biomass can be converted by processes such as combustion, gasification or pyrolysis. In gasification, for instance, the main product is a mixture of gases that can be utilised either as a gaseous fuel to be burnt in a gas turbine or engine for the production of heat or electricity, or both simultaneously. A second option for using this raw gas is for the synthesis of liquid fuels such as methanol, hydrogen and hydrocarbons. Similarly, biochemical biomass can be converted into biogas by enzymes in an anoxic environment (in the absence of oxygen).

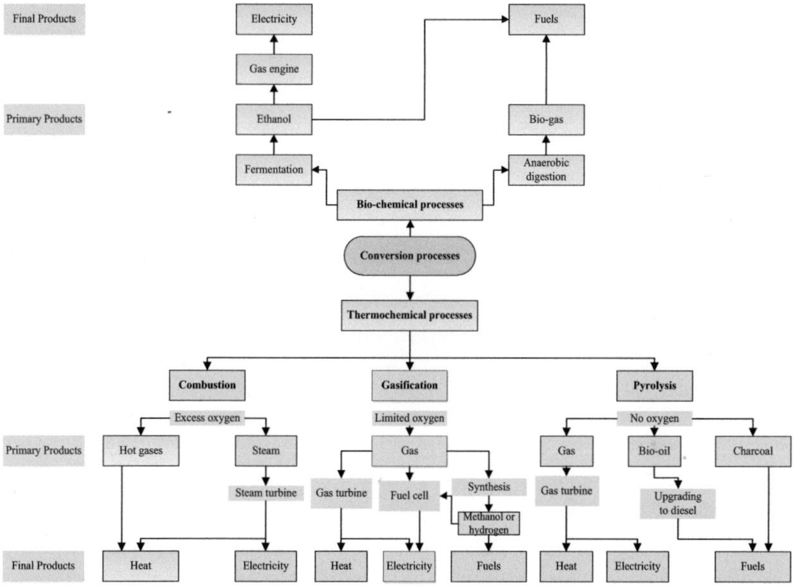

Figure 3.11: Biomass conversion processes and their products (adapted from Akhtari *et al.* 2014).

Biogas, a mixture of methane and carbon dioxide of variable composition, can be either directly burnt in an engine for the simultaneous production of heat and electricity or utilised for the generation of a gaseous fuel via upgrading. The production of end-products (i.e. electricity, heat or fuels) can be carried out through conversion routes, depending on the physicochemical process which the conversion is based on (i.e. combustion, gasification, etc.), which at the same time is conditioned by the sort of feedstock (biomass) to be converted (cf. Figure 3.11). For this reason the processes are normally grouped into thermochemical conversion processes, those based on the use of heat for the promotion of the reaction of the biomass, biochemical conversion processes, those based on the action of microorganisms for the conversion, and the physical conversion processes, those in which no chemical reaction takes place (at least at that stage). One distinguishing characteristic of biomass conversion is its flexibility, i.e. the possibility of converting biomass into different intermediates or end-products depending on what the most desirable option or product is. The final decision on the end-product will depend on numerous factors and is, in most cases market-driven.

3.2.3.1 Multiple uses of biomass

Unlike other renewable resources (i.e. wind, photovoltaic or tidal energy) that can only displace power generation, biomass has multiple uses that may not only imply inter-competition between biomass consumers but can also indirectly trigger inter-competition for land, water and the hegemony on setting land use criteria (Spangenberg 2007). Hoogwijk *et al.* (2005) schematically depict the mutual influence of land use and production of food, materials and energy as Figure 3.12 shows. At the global scale and in a simplified representation, this scheme illustrates the mutual influences between various primary sources and chains of supply for biomass. For instance, food processing waste can be used as fodder (dotted line), material production or energy. Although not explicitly shown, it is intuitively clear that there is an interrelated effect of food, material and energy consumption on land use as well as a correlation between the demand of biomass for one given application and its effect on others.

The relationship between the production of chemicals (starch, oil, sugars), construction materials (wood, plywood, etc.) and bioenergy stand out from this web of interconnections; land for human nutrition, fodder for animals and energy crops involves another group of complex interactions. This particularity of biomass may bring an additional level of difficulty into calculating or forecasting the share of new biomass-derived products in a particular market (e.g. electricity, heat) and its effect on other markets (e.g. edible goods, fertilisers, materials, etc.) as well as impacts on ecosystems.

Based principally on these facts, Spangenberg (2007) casts doubt on the possibility that bioenergy may provide a meaningful contribution in the future unless radical changes in patterns of energy consumption were made at system level.

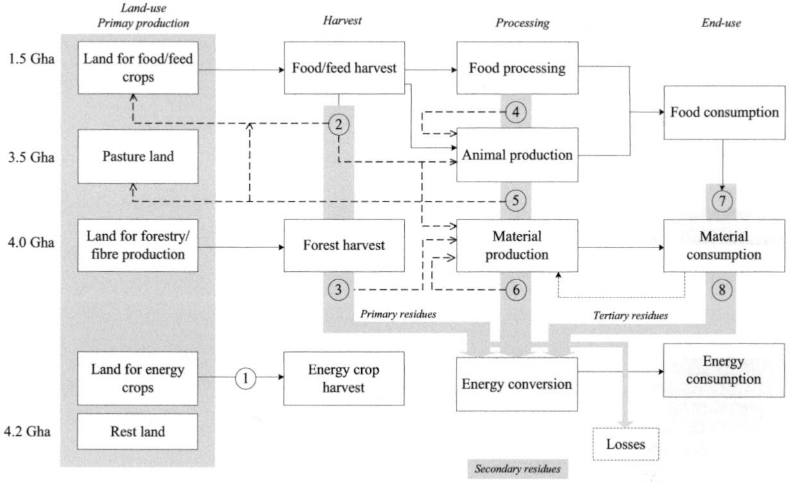

Figure 3.12: Land use and some flows of biomass at global scale (adapted from Hoogwijk *et al.* 2005).

3.2.3.2 Biomass conversion technologies

This section gives a brief outlook of six technologies that RA3 assessed. It contains a basic description of the functional principle of the technology, level of its commercialisation and type of biomass that it can process. As the economic characterisation of each technology in terms of its economics is beyond the scope of this succinct report, more detailed information can be found in the full report that DFIU produced[1].

[1] Available at the project website www.oui-biomasse.info

Anaerobic digestion

Anaerobic digestion is widely thought to be one of the most energy-efficient and environmentally friendly technologies for bioenergy production (Weiland 2010; OECD 2010). Moreover, it has outstanding environmental, technical as well as social advantages that even make arguable its direct economic comparison with traditional fossil-based processes for the generation of energy (Berglund 2006; Börjesson & Berlung 2006). The basic technology of anaerobic digestion converts organic biomass into methane and carbon dioxide, biogas, in a variable composition normally ranging from 50% to 65% in methane (molar base).

The anaerobic digestion process occurs in a reactor, usually a so-called digester, where biomass undergoes reaction by the action of different consortia of microorganisms in an anoxic media, i.e. in absence of oxygen. These microorganisms break biodegradable solids and soluble matter down, thus producing biogas. The anaerobic digestion process can be carried out by using either wet-fermentation or dry-fermentation technologies (Deublein & Steinhauser 2011; Weiland 2010). According to Karellas *et al.* (2010), the former operates with a concentration (measured as dry solid) lower than 20%, whereas the latter is adequate for a substrate with water content lower than 85%.

Combustion

The combustion of biomass in air is used to turn the energy chemically stored in biomass into heat, which can be subsequently used for the generation of electricity, mechanical work or simply for heating (Klass 1999). It is the most straightforward way to utilise biomass as it implies the complete oxidation of carbonaceous substances. The technology available for biomass combustion covers a wide range of power and applications, and it consists, for instance, of stoves, furnaces, boilers, heating plants, and large-scale power generation facilities (Widell 2013). To generate electricity at industrial scale, large boilers are commonly coupled with steam turbines, generators and other auxiliary equipment for energy generation, which is the dominant technology.

In spite of being a state-of-the-art application, the combustion of biomass is a complex process from a technological point of view (Turns 2000). Biomass combustion involves multiple chemical reactions and advanced abatement systems for reducing air emissions from flue gas are included in modern biomass combustion systems.

Gasification

Gasification is a thermochemical process in which a feedstock is partially oxidised at relatively high temperatures (500-1 400°C) using a restricted amount of oxygen. The pressure of operation can vary from atmospheric to 33 bar (IRENA 2012). The substance that contains oxygen and is used to carry out gasification is called the gasifying agent, and the product of the gasification is called synthesis gas, or simply syngas. Although the gasifying agents can have considerable influence on the gasification reaction, syngas is made up mainly of carbon dioxide, molecular hydrogen, methane and a mixture of hydrocarbon. The quantity and composition of agents are dependent on the gasification environment.

The equipment where gasification reactions take place is normally called a gasifier. Gasifiers can be classified according to different criteria such as type of gasifying agent (air, oxygen or stem), working pressure (pressurised or atmospheric) and way of transferring heat to the process (allothermal or autothermal), the most common means of classification being the gas-solid contacting mode (Tremel, *et al.,* 2013; IRENA 2012). Following this convention, gasifiers can be grouped into three main families: fixed bed (downdraft or updraft bed), fluidised bed (bubbling fluidised bed or circulating fluidised bed) and entrained flow.

Fermentation

Fermentation is an industrial technology for the production of bioethanol from sugar or starch-rich substances (Murphy & Power 2008). Fermentation starts with the mechanical pre-treatment of feedstock, which helps enzymes convert starch into sugars and then yeast convert sugar into ethanol. Finally, the purification of ethanol is carried out by distillation, a step which requires high energy consumption to reach a bioethanol concentration higher than 99% (Murphy and Power 2008). The by-products of these processes are denominated distiller dry grains and solubles (DDGS), and they are commonly used as fodder for livestock (Power *et al.* 2008).

Bioethanol is used mainly as a substitute for gasoline; however, there are significant differences between bioethanol and gasoline. For instance, bioethanol has a higher concentration of oxygen (34%) than gasoline (0%), an almost unlimited solubility in water and lower heating value. Therefore, bioethanol cannot be directly used in conventional engines; hence its use in blends.

The production of bioethanol from energy crops such as sugar beet, corn and wheat is a mature technology. Nevertheless, the production of bioethanol from lignin-rich biomass is still not commercial (Gnansounou & Dauriat 2010) although there are numerous prototype projects that aim to develop this novel technology.

Biomass-to-liquid

Biomass-to-liquid is a process for the production of liquid fuels from the thermochemical conversion of lignocellulosic biomass (Trippe *et al.*, 2011). This process has the distinguishing characteristic of the densification of biomass with the use of multiple pyrolysis plants to make its transportation more economically competitive. After the decentralised pyrolysis of biomass, the produced bio-slurry is carried to a central plant and then gasified for the production of syngas. Due to issues of economies of scale, the gasification of bio-slurry has to be carried out on a large scale. Afterwards, synthesis takes place for the production of synthetic fuels or even chemicals (Haro *et al.*, 2013).

Waste-to-energy

Waste-to-energy, previously known as incineration, primarily uses municipal solid waste (MSW) as a fuel, with or without sorting. Waste-to-energy has exhibited a noteworthy improvement in performance in recent years, with the integration of enhanced abatement control of pollutants; therefore, it has become a clean technology for energy generation and an integral part of advanced municipal solid waste management systems (Bidart *et al.*, 2013). Although there are a variety of waste-to-energy technologies, the most common works with vibrating grates.

The use of waste-to-energy technologies involves substantial investment and high cost of operation. It is an alternative of processing municipal solid waste that has to be incorporated into a global framework to process municipal solid waste that ought to prioritise recycling, composting, sorting of organic and inorganic matter, and finally landfilling (Ludwig *et al.,* 2010).

3.2.3.3 Concluding remarks

Biomass comprises a wide range of feedstock such as municipal solid waste, manure, agricultural residues, energy crops and forestry residues. Its utilisation has considerable advantages over fossil-based fuels; for instance, it is widely distributed, so it can be locally employed and its utilisation can lead to significant reduction of carbon dioxide emissions. Nevertheless, the utilisation of biomass is associated with problems principally related to its significantly lower energy density, which makes its transportation, storage and pre-treatment costly. Moreover, its markedly different chemical composition, when compared with coal or hydrocarbons, poses significant challenges for its conversion into useful forms of energy.

A range of technologies has been developed for the utilisation of biomass although direct combustion is still the dominant means of biomass conversion. Figure 3.13 shows in a diagrammatic form the stage of development for some biomass conversion technologies, from those that are in a basic and applied R&D stage to those that are fully commercial.

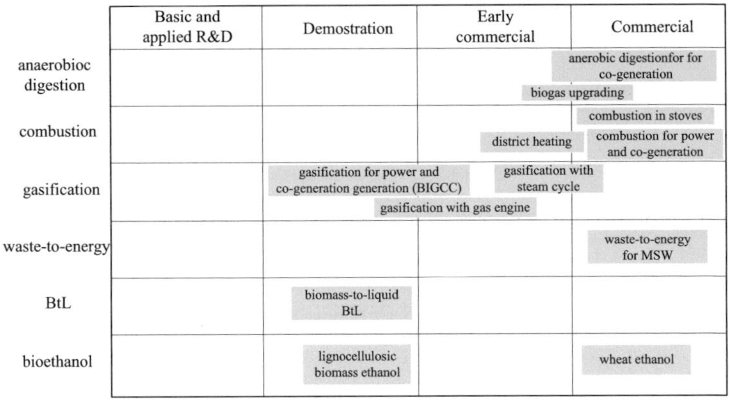

Figure 3.13: Schematic representation of the stage of development for some biomass conversion technologies assessed in 'OUI Biomasse' Project (adapted from IEA 2012).

While anaerobic digestion and combustion are proven technologies, gasification is in an early commercial or demonstration stage. Biomass Integrated Gasification Combined Cycle (BIGCC) has shown advantages in terms of efficiency in comparison with combustion; however, aspects related to reliability and cost of biomass pre-treatment have to be solved. Similarly, on one hand the production of bioethanol by processing wheat is a commercial process widely employed on a large scale. On the other hand, the processing of lignocellulosic biomass for ethanol production is still at demonstration stage.

Waste-to-energy based on conventional combustion is fully commercial and has become a concrete option for the treatment of municipal solid waste. The incorporation of advanced abatement systems makes it feasible to fulfill tough environmental regulations. Finally, the BtL technology is a process in demonstration phase; its innovative approach is based on the decentralised densification of biomass by pyrolysis, which makes it unique in the field of biomass-to-fuel concepts. Furthermore, the inclusion of gasification of biomass on a large scale is an intermediate step that would facilitate the adoption of a biorefinery approach for not only the production of synthetic fuels but also as platform for the synthesis of chemicals.

3.2.4 Microbiological energy valorisation of grape marc

Stéphane Vuilleumier, *Muhammad Farhan Ul Haque, Christelle Gruffaz, Ludovic Besaury*

Génétique Moléculaire, Génomique, Microbiologie (GMGM), Université de Strasbourg, Strasbourg, France.

3.2.4.1 Context of the demostration study

Disposal of grape marc is an unresolved issue in wine production. If untreated, marc is associated with various environmental issues ranging from pollution, in particular of groundwater and potential eutrophication, to foul odors. Traditionally, grape marc has been used as compost or distilled to produce alcohol. Recently, grape spirit oversupply has rendered distillation economically unattractive, and alternative valorisation strategies of marc, including for bioenergy production, have been increasingly considered (Fabbri *et al.* 2015). However, fundamentally unresolved issues remain as to the biogas and biomethane production potential of marc, and its dependence on grape variety, marc conditioning, and associated microbiological status of grape marc. Regarding the latter, recent developments in molecular microbial ecology make it possible to characterise the microbial diversity associated with grape marc in unprecedented detail.

In this context, partner GMGM at Université de Strasbourg chose to address the following two main questions in an exploratory demonstration study within 'OUI Biomasse':

What is the influence of grape variety and marc conditioning on biogas and biomethane production from grape marc?

Is the microbial composition of grape marc correlated with biogas and biomethane production?

The main results obtained in this study are briefly summarised in the following.

3.2.4.2 Fermentation experiments

The grape marc samples investigated in this study were the same as those used by laboratory GRE in their investigations on the technical application of grape residues (Section 2.2). The methanogenic potential of grape marc was evaluated by two different methods and at two different scales, by EIFER working as a subcontractor on the project. The 100 ml Hohenheim Biogas-yield Test (HBT) reactor (Lang 2012) was applied on finely ground marc samples for initial screening of marc of different grape varieties and conditioned in different ways. Scale-up was also performed in 2L-bioreactors with coarsely ground samples to better mimick operating conditions in an industrial methanisation plant.

Screening of biogas and biomethane production potential from grape marc

The HBT (Hohenheim Biogas-yield Test) reactor used in the initial screening allowed to evaluate the effect of grape variety and conditioning on production of biogas and biomethane (Figure 3.14). The patented HBT test consists of a thermostatted (37°C) rotating device which can accommodate up to 129 100ml graduated syringes disposed horizontally, allowing for efficient mixing of marc substrate (1-2 g per syringe) and manure inoculum. Gas production is quantified and methane content is determined using an infrared sensor. Using this setup, samples best suited for methane production could be selected after two weeks.

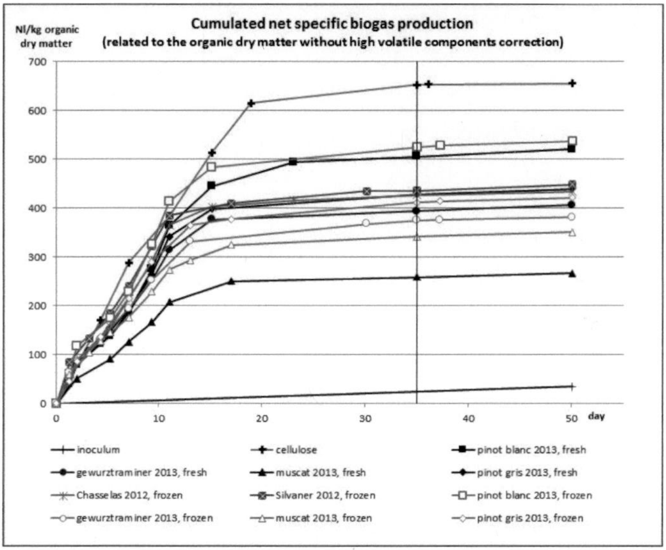

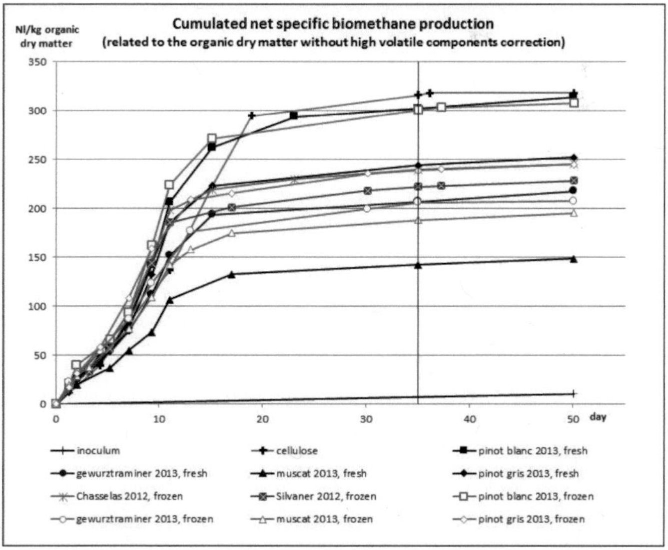

Figure 3.14: Cumulated net specific production of biogas and biomethane through time as a function of grape marc type and conditioning.

The mean values obtained were in the range of those reported in the literature. Although methanogenic potential was not correlated with the content of volatile organic compounds or with sugar, an interesting effect of grape variety was observed (Table 3.7).

It appears that marc of the Pinot blanc variety performs best in terms of methane production, both for fresh and frozen samples. Mean biogas production values (for a total of 10 different samples) were of 410 Nl/kg organic dry matter, with a mean content of approximately 56% (i.e. 229 Nl/kg organic dry matter).

Different grape varieties were also differentially affected by the mode of conservation. Specifically, Pinot blanc was least affected, and Muscat most affected, by conditioning. Nevertheless, the resulting variations in production of biogas and biomethane were quite minor overall.

Table 3.7: Specific mean methanogenic production at 35 days.

	Specific production (Nl/kg organic dry matter)		Reactor type	Methane in biogas (%)
	Biogas	Methane		
Pinot blanc 2013, fresh	504	302	HBT	60
Pinot blanc 2013, frozen	525	301	HBT	57
Pinot blanc 2013, frozen	507	189	2L	37
Gewürztraminer 2013, fresh	393	207	HBT	53
Gewürztraminer 2013, fresh	448	96	2L	21
Gewürztraminer 2013, frozen	375	206	HBT	55
Gewürztraminer 2013, frozen	393	139	2L	35
Muscat 2013, fresh	295	143	HBT	55
Muscat 2013, frozen	342	188	HBT	55
Pinot gris 2013, fresh	428	244	HBT	57
Pinot gris 2013, frozen	412	239	HBT	58
Pinot gris 2013, frozen	507	189	2L	37
Chasselas 2012, frozen	426	240	HBT	56
Silvaner 2012, frozen	436	222	HBT	51

Initial attempts at scale-up of biogas production from grape marc

Following the initial screening of biogas production using the HBT reactor, frozen samples of the two best-performing marcs (Pinot blanc and Pinot gris) were analysed in 2L bioreactors, together Gewürztraminer marc under both fresh and frozen forms (Table 3.7).

The 2L-bioreactor setups involved modified GLS 80 Duran (Schott) glass reactors, thermostatted at 37°C and equipped with a blade agitator operating at 25 rpm. Biogas formed in the reactor is evacuated and analysed by a MilliGascounter and microscale gas chromatography. Approximately 50-100 g of marc was used per experiment. Again, the Pinot Blanc variety displayed the highest methanogenic potential. No clear trend was observed in the values obtained in HBT and 2L reactors, which overall gave similar results. Also and as observed with the HBT reactors, freezing did not significantly affect methanogenic potential. One important factor affecting performance is the extent of grinding (<10 mm in 2L reactors, compared to <1 mm fragments in HBT reactors). The coarser grinding in 2L reactors is a possible explanation between the poorer reproducibility between replicate runs on the same marc compared to HBT experiments in this study. This, together with the related issue of the proportion of solid material in marc preparations, is most likely the crucial issue to consider in large-scale biomethane production.

3.2.4.3 Bacterial and archaeal diversity analysis

All grape marc samples investigated for their biogas forming potential were analysed in terms of the diversity of their microbial flora. Not only Bacteria but also Archaea were investigated, since Archaea comprise the only known methanogenic micro-organisms, while the majority of fermenting microbes belong to Bacteria. To this end, total DNA was isolated from frozen marc samples and a diagnostic region of the universally conserved 16S ribosomal RNA gene was applied for taxonomical analysis of the microbial diversity in all marc samples following high-throughput DNA sequencing. To our knowledge, this is one of the very first such analyses to be reported on grape marc (see e.g. Maragkoudakis *et al.*, 2013; Campanaro *et al.*, 2014), and perhaps the first performed on grape marc following its valorisation through biogas production.

Procaryotic diversity of grape marc samples

Total DNA was extracted from thawed grape marc samples before and after biogas production as well as from the inoculum used to initiate biogas production (approx. 250 mg) using a commercial DNA extraction kit (PowerSoil®, MoBio) chosen to give the best compromise in terms of DNA yield and purity (0.12-3.10 g DNA, mean 1.68±1.00 g DNA). The hypervariable V4-V5 region of the 16S rRNA gene was amplified using oligonucleotide primers GTGYCAGCMGCCGCGGTA and CCCCGYCAATTCMTTTRAGT by the polymerase chain reaction, and sequenced using Illumina® MiSeq high-throughput sequencing technology, through the services of a commercial provider.

Taxonomic analysis of the microbial diversity in all samples was performed using the Mothur pipeline (Kozich *et al.*, 2013) allowing the relative proportion of Bacteria and Archaea, as well as that of the major taxa within these two domains to be determined at the family level. The main initial results from this analysis of 2 061 597 sequences (128 674 unique sequences) in of 362-418 nucleotides (mean: 374 nucleotides) in total for 20 investigated samples are summarised in Figure 3.15 and Table 3.8.

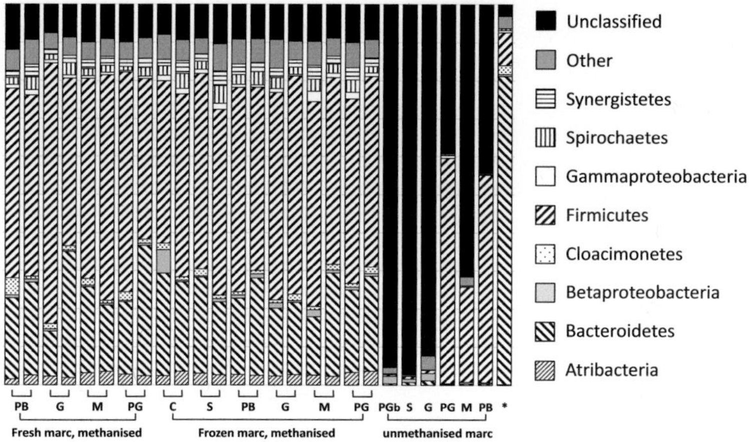

Figure 3.15: Relative abundance of major bacterial and archaeal phyla in grape marc samples. Two replicate DNA extracts per inoculated grape marc were sequenced. PB: Pinot blanc. G: Gewürztraminer; M: Muscat; PG: Pinot gris; C: Chasselas (2012); S: Sylvaner (2012); PGb: Pinot gris (2012); *: manure inoculum.

No large variation in the major bacterial families were detected as a function of grape variety or conditioning. Far fewer sequences were retrieved from uninoculated grape marc samples, possibly due to higher inhibition of DNA amplification by humic acid inhibitors in such woody, unprocessed samples. They also presented a very different microbial composition than grape marc after use for biogas production, as again did the inoculum used to initiate biogas formation. In all cases, Archaeal families and thus methanogenic organisms, represented only a minor fraction of the total prokaryotic composition not only in uninoculated sample, but also in marc samples after biogas production and also, perhaps more surprisingly, in the inoculum itself (Table 3.8). Worthy of note, the proportion of archaeal sequences was highest in Pinot blanc, paralleling the observed highest methane production with this grape variety (Table 3.7). At the potentially more resolutive family level, however, the relative proportion of the generally most abundant archaeal *Methanomassiliicoccaceae* family was relatively high. in Pinot blanc, but highest in Sylvaner grape marc, found here to be associated with a lower methanogenic potential (Table 3.8).

Table 3.8: Total number of sequence and relative abundance of Archaea and the major families of Bacteria (unclassified Bacteroidetes) and Archaea (Methanomassiliicoccaceae) in grape marc samples.

Sample	Bacteria		Archaea	
	Total retrieved sequences	Bacteroidetes (%)	Total (%)	Methanomassiliicoccaceae (%)
Inoculated with manure [a]				
Pinot blanc 2013, fresh	120 303	18.29	0.73	0.28
Gewürztraminer 2013, fresh	106 479	20.18	0.50	0.18
Muscat 2013, fresh	76 010	17.33	0.48	0.19
Pinot gris 2013, fresh	76 794	23.41	0.48	0.17
Chasselas 2012, frozen	49 950	19.51	0.41	0.19
Sylvaner 2012, frozen	79 030	18.56	0.63	0.33
Pinot blanc 2013, frozen	86 313	18.55	0.67	0.24
Gewürztraminer 2013, frozen	68 088	15.26	0.58	0.19
Muscat 2013, frozen	140 986	18.32	0.36	0.11
Pinot gris 2013, frozen	78 523	19.29	0.57	0.23
Uninoculated				
Pinot gris 2012	8 008	0.00	ND[b]	ND
Sylvaner 2012	7 926	0.23	ND	ND
Gewürztraminer 2013	6 260	0.14	ND	ND
Pinot gris 2013	6 576	0.09	ND	ND
Muscat 2013	6 942	0.02	ND	ND
Pinot blanc 2013	16 920	0.33	ND	ND
Inoculum				
Manure	36 253	78.60	0.03	0.01

[a] Total sequences of two replicates, and mean percentages of two replicates.
[b] Not detected.

A detailed biostatistical analysis of the obtained data at the finer taxonomic level of individual bacterial genera (361 archaeal and 28 534 bacterial phylotypes) is currently underway, in order to evaluate whether specific bacterial or archaeal phylotypes at the even more resolutive genus level also correlate with methane production and/or grape type or grape marc conditioning and could potentially be used as bioindicators of methanogenic potential of grape marc.

3.2.4.4 Conclusions and perspectives

Grape variety appears to be a key factor in biogas and biomethane production from grape marc. Up to two-fold differences in methane yield were observed. In both HBT and 2L reactor types, the Pinot blanc grape variety displayed the highest methanogenic potential among the grape varieties investigated in our study. The finer grinding achievable for experiments with the HBT reactor experiments and a lower proportion of dry parts also contributed to a higher methane production. In contrast, the mode of conservation of grape marc (i.e. fresh versus frozen) appears to have only a minor impact, although clear trends could not be made out.

With regard to the microbial composition of marc grape, major differences between samples at the phylum level were not evident beyond those expected between marc samples inoculated or not with manure. Archaea and methanogens also make out a very minor proportion of the total prokaryotic diversity in all marc samples. In the future, a higher resolution taxonomic analysis of the data obtained will show whether specific phylotypes are associated with methanogenic potential and thereby serve for bioindication purposes to define the grape marc preparations which are the best suited for valorisation through biomethane production.

3.2.5 References section 3.2

References subsection 3.2.1: Catalysts for syngas methanation

Jensen, JH, Poulsen, JM, Andersen, NU 2011, 'From coal to clean energy', *Nitrogen+Syngas*, vol. 310, Available from: <http://www.topsoe.com/sites/default/files/topsoe_from_coal_to_clean_energy_nitrogen_syngas_march_april_2011.pdf>.

Kopyscinski, J, Schildhauer, TJ, Biollaz, SMA 2010, 'Production of synthetic natural gas (SNG) from coal and dry biomass - A technology review from 1950 to 2009', *Fuel,* vol. 89, no. 8, pp. 1763-1783.

Nguyen, TTM, Wissing, L, Skjøth-Rasmussen, MS 2013, 'High temperature methanation: Catalyst considerations', *Catalysis Today,* vol. 215, pp. 233-238.

Ocampo, F, Louis, B, Kiwi-Minsker, L, Roger, AC 2011, 'Effect of Ce/Zr composition and noble metal promotion on nickel based $Ce_xZr_{1-x}O_2$ catalysts for carbon dioxide methanation', *Applied Catalysis A: General*, vol. 392, pp. 184-191.

Rostrup-Nielsen, JR, Pedersen, K, Sehested, J 2007, 'High temperature methanation: Sintering and structure sensitivity', *Applied Catalysis A: General*, vol. 330, pp. 134-138.

References subsection 3.2.2: Investigation of technical usability of wine residue

Brillard, A, Brilhac, JF, Valente, M 2015, 'Modelization of the grape marc pyrolysis and combustion based on an extended independent parallel reaction and determination of the optimal kinetic constant', *Computational & Applied Mathematics*, published on-line 15 feb. 2015.

Celma, AR, Rojas, S, López-Rodríguez, F 2007, 'Waste-to-energy possibilities for industrial olive and grape by-products in Extremadura', *Biomass and Bioenergy*, vol. 31, pp. 522–534.

DIN CERTCO, DIN EN 13229 2008, 'Certification Scheme – Fireplace insert appliances and cassettes fired by solid fuels with low-pollution combustion'.

Dorge, S, Jeguirim, M, Trouvé, G 2011, 'Thermal degradation of Miscanthus pellets: kinetics and aerosols characterization', *Waste Biomass Valorisation*, vol. 2, pp. 149–155.

Favez, O, Cachier, H, Sciare, J, Sarda-Estève, R, Martinon, L 2009, 'Evidence for a significant contribution of wood burning aerosols to PM2.5 during the winter season in Paris, France', *Atmospheric Environment*, vol. 43, no. 22–23, pp. 3640-3644.

Johansson, LS, Leckner, B, Gustavsson, L, Cooper, D, Tullin, C, Potter, A 2004, 'Emission characteristics of modern and old-type residential boilers fired with wood logs and wood pellets', *Atmospheric Environment*, vol. 38, pp. 4183–4195.

Marjamäki, M, Ristimäki, J, Virtanen, A, Moisio, M, Luoma, R, Keskinen, J 2000, 'Testing porous metal as a collection substrate in ELPI', *Journal of Aerosol Science*, European Aerosol Conference 2000, vol. 31, Supplement 1, pp. 76–77.

NF EN 303-5 2012, 'Heating boilers – Part 5: Heating boilers for solid fuels, manually and automatically stoked, nominal heat output of up to 500 kW – Terminology, requirements, testing and marking'. NF EN 303-5.

NF EN 304/A1 1998, 'Heating boilers — Test code for heating boilers for atomizing oil burners'. NF EN 304/A1.

NF EN 12809/A1 2005, 'Residential independent boilers fired by solid fuel – Nominal heat output up to 50 kW – Requirements and test methods'. NF EN 12809/A1.

Saidur, R, Abdelaziz, EA, Demirbas, A, Hossain, MS, Mekhilef, S 2011, 'A review on biomass as a fuel for boilers', *Renewable and Sustainable Energy Reviews,* vol. 15 no. 5, pp- 2262-2289.

Toscano, G, Riva, G, Duca, D, Pedretti, E, Corinaldesi, F, Rossini, G 2013, 'Analysis of the characteristics of the residues of the wine production chain finalized to their industrial and energy recovery', *Biomass and Bioenergy*, vol. 55, pp. 260–267.

Valente, M, Brillard, A, Schönnenbeck, C, Brilhac, J 2015, 'Investigation of grape marc combustion using thermogravimetric analysis. Kinetic modeling using an extended independent parallel reaction (EIPR)', *Fuel Processing Technology*, vol.131, pp. 297‐303.

Valente, M, Trouvé, G, Schonnenbeck, C, Brilhac, J 2014, 'Emission factor of gas and particulate matter during the energy recovery of grape marc in a domestic boiler'. In: Conference Proceedings of the 1st International Conference on Atmospheric Dust - DUST 2014.

Verma, V, Bram, S, Gauthier, G, De Ruyck, J 2011, 'Evaluation of the performance of a multi-fuel domestic boiler with respect to the existing European standard and quality labels: Part-1', *Biomass and Bioenergy*, vol. 35, pp. 80–89.

References sub-section 3.2.3: Biomass conversion technologies

Berglun, M 2006, '*Biogas production from a systems analytical perspective'*, Lund University. Faculty of Engineering. Department of Technology and Society. Environmental and Energy Systmes Studies, Sweden. Doctoral dissertation.

Bidart, C, Fröhling, M, Schultmann, F 2013, 'Municipal solid waste and production of substitute natural gas and electricity as energy alternatives', *Applied Thermal Engineering*, vol. 51, pp. 1107-1115.

Börjesson, P, Berglund, M 2006, 'Environmental systens analysis of biogas systems – Part I: Fuel-cycle emissions', *Biomass and Bioenergy*, vol. 30, pp. 469-485.

Deublein, D, Steinhauser, A 2011, 'Biogas from waste and renewable resources', 2nd ed., Wiley-VCH Verlag GmbH & Co. KGaA, Weinheim.

Gnansounou, E, Dauriat, A 2010, Techno-economic analysis of lignocellulosic ethanol, A review, *Bioresource Technology*, vol.101, pp. 4980-4991.

Haro, P, Trippe, F, Stahl, R, Henrich, E 2013, 'Bio-syngas to gasoline and olefins via DME – A comprehensive techno-economic assessment', *Applied Energy*, vol. 108, pp. 54-65.

Hoogwijk, M, Faaij, A, Eickhout, B, de Vries, B, Turkenburg, W 2005, 'Potential of biomass energy out to 2100, for four IPCC SRES land-use scenarios', *Biomass and Bioenergy*, vol. 29, pp. 225-257.

IRENA (International Renewable Energy Agency) 2012, *Renewable energy technologies: cost analysis series*, vol. 1, issue 1/5, Available from: <www.irena.org/DocumentDownloads/Publications/RE_Technologies_Cost_Analysis-BIOMASS.pdf>. [05.08.2015].

IEA (International Energy Agency) 2012, *Technology roadmap. Bioenergy for heat and power*, Paris, France, Available from: <www.iea.org/publications/freepublications/>. [06.08.2015].

Karellas, S, Boukis, I, Kontopoulos, G 2010, 'Development of an investment decision toll for biogas production from agricultural waste', *Renewable and Sustainable Energy Reviews*, vol. 14, pp. 1273-1282.

Klass, D 1998, *Biomass for renewable energy, fuels, and chemicals*, Academic Press Inc, London.

Ludwig, C, Hellweg, S, Stucki, S 2013, *Municipal solid waste management. Strategies and technologies for sustainable solutions*, 1st ed., Springer-Verlag, Berlin-Heidelberg.

Murphy, J, Power, N 2008, 'How can we improve the energy balance of ethanol production from wheat?', *Fuel*, vol. 87, pp. 1799-1806.

OECD 2010, *Bioheat, biopower and biogas. Development and implications for agriculture*, OECD, Paris.

Power, N, Murphy, J, McKeogh, E 2008, 'What crop rotation will provide optimal first-generation ethanol production in Ireland, from technical and economic perspectives?', *Renewable Energy*, vol. 33, pp. 1444-1454.

Spangenberg, J 2013, 'Biomass or biomess? The promises and limits of bioenergy'. In: Ruppert, H, Kappas, M, Ibendorf, J (ed.), *Sustainable bioenergy production - an integrated approach*, Springer, London.

Tremel, A, Becherer, Fendt, S, Gaderer, M, Spliehoff, H 2013, 'Performance of entrained flow and fluidised bed biomass gasifiers on different scales', *Energy Conversion and Management*, vol. 69, pp. 95-106.

Trippe, F, Fröhling, M, Schultmann, F, Stahl, R, Henrich, E 2011, 'Techno-economic assessment of gasification as a process step within biomass-to-liquid fuel and chemical production', *Fuel Processing Technology*, vol. 923, pp. 2169-2184.

Turns, S 2000, *An introduction to combustion. Concepts and Applications*, 2nd ed., McGraw-Hill, Singapure.

Weiland, P 2010, 'Biogas production: current state and perspectives', *Applied Microbiology and Biotechnology*, vol. 85, pp. 849-860.

Widell, H 2013, 'Industrial-scale biomass combustion plants: engineering issues and operations'. In: Rosendahl, L (ed.), *Biomass combustion science, technology and engineering*, Woodhead Publishing, Cambridge.

References subsection 3.2.4: Microbiological energy valorisation of grape marc

Campanaro, S, Treu, L, Vendramin, V, Bovo, B, Giacomini, A, Corich, V 2014, 'Metagenomic analysis of the microbial community in fermented grape march reveals that Lactobacillus fabifermentans is one of the dominant species; insights into its genome structure', *Applied Microbiology and Biotechnology*, vol. 98, pp. 6015-6037.

Fabbri, A, Bonifazi, G & Serranti, S 2015, 'Micro-scale valorization of grape marcs in winery production plants', *Waste Management*, vol. 36, pp. 156-165.

Kozich, JJ, Westcott, SL, Baxter, NT, Highlander, SK & Schlossn PD 2013, 'Development of a dual-index sequencing strategy and curation pipeline for analyzing amplicon sequence data on the MiSeq Illumina sequencing platform', *Applied and Environmental Microbiology*, vol. 79, pp. 5112-5120.

Lang, S 2012, *Energetische Verwertung weinbaulicher Nebenprodukte. Die Rolle der nachhaltigen Landwirtschaft für die Energiewende*, Agroscience IFA.

Maragkoudakis, PA, Nardi, T, Bovo, B, D'Andrea, M, Howell, KS, Giacomini, A & Corich, V 2013, 'Biodiversity, dynamics and ecology of bacterial community during grape march storage for the production of grappa', *International Journal of Food Microbiology*, vol. 162, pp. 143-151.

4 Perspectives of future energetic biomass production and use

Nadège Blond[1], Mathieu Bosansky[2], Joëlle Duplay[3], Yaxian Hu[4], Martin Knapp[5], Rolf Meyer[5], Ann-Kathrin Müller[6], Alexandra Pehle[5], Cyril Pallarès[2], Jean-Luc Ponche[1,7], Gaëtana Quaranta[3], Andreas Rudi[6], Wissal Selmi[1], Gerhard Schäfer[3], Frank Schultmann[6], Jérémy Stark[1], Volker Stelzer[5]

[1] Laboratoire Image, Ville, Environnement (LIVE), Université de Strasbourg/ CNRS, Strasbourg, France.

[2] Association pour la Surveillance de la Pollution Atmosphérique (ASPA), Strasbourg, France.

[3] Laboratoire d'HYdrologie et de GEochimie de Strasbourg (LHyGeS), Université de Strasbourg/ CNRS, Strasbourg, France.

[4] Departement Umweltwissenschaften (DUW), Universität Basel, Basel, Schwitzerland.

[5] Institute for Technology Assessment and Systems Analysis (ITAS), Karlsruhe Institute of Technology (KIT), Karlsruhe, Germany.

[6] French-German Institute for Environmental Research (DFIU), Karlsruhe Institute of Technology (KIT), Karlsruhe, Germany.

[7] Institut de Chimie et Procédés pour l'Energie, l'Environnement et la Santé (ICPEES), Université de Strasbourg/CNRS, Strasbourg, France.

4.1 Biomass scenarios and their results

***Rolf Meyer**, Alexandra Pehle, and Martin Knapp*

Institute for Technology Assessment and Systems Analysis (ITAS), Karlsruhe Institute of Technology (KIT), Karlsruhe, Germany.

4.1.1 Introduction

Main target of creating and analysing the scenarios within the framework of the project OUI Biomasse is to enable „a stakeholder dialogue about the best elements for a regional roadmap to a sustainable utilisation of biomass“ (Knapp et al. 2014). The scenarios map possible diverging developments of future biomass production and use in the Upper Rhine Region (URR) to achieve this objective.

This section outlines the scenario approach, presents the elaborated scenarios and main results of the scenario analysis. The scenario analysis is based on the data sources and assessment approaches for agricultural, forestry land use and biomass production, as well as for waste biomass as described in section 3.1.

4.1.2 Scenario approach

The scenarios represent possible developments of future biomass production in agriculture and forestry and available biomass in the waste sector of the URR, especially in regard to their potential use for bioenergy production. A *medium-term timeframe*, until the year 2030, is chosen for the scenarios.

Exploratory scenarios were developed. They are based on relevant national and European framing conditions, which, in the field of bioenergy, influence future production and use of biomass in the URR. The scenario construction is based on four *main drivers*, that were appointed after a multiphase process of literature research and discussions, through which important and relevant factors were identified (Knapp et al. 2014). The drivers are:

- Energy policies
- Agricultural policies
- Nature conservation policies
- Bioenergy technologies

Other future developments such as energy prices or the economic development in the URR are also uncertain and are potentially influencing the bioenergy use. But they will have less influence than the selected drivers. Additionally, the number of drivers needs to be restricted for a plausible scenario construction.

The four drivers listed above are broad policy fields which include different areas of action with relevance for the regional bioenergy production. Therefore, each of these drivers is subdivided into several *policy areas or influencing factors* that are relevant for different areas of bioenergy use and for limitations of possible utilisations, respectively.

The scenarios are determined by scenario specific assumptions regarding both, the four drivers and each influencing factor (see section 4.1.3). As a matter of principle, with respect to the trinational region, it is assumed that in each scenario the same policies and developments are valid for the two EU-countries France and Germany, as well as for Switzerland, while respecting national differences (e.g., there will be no biofuel quota in Switzerland in the future).

The demand for energetic biomass usage in 2030, with the corresponding need for biomass and the need for areas, is derived from the scenario assumptions on future energy policies. The scenario assumptions on future agricultural and nature conservation policies influence the available area for biomass production and the kind of production.

In the fields of renewable electricity and heat, the currently existing energetic biomass usage in the region is kept up as a baseline until 2030. The reasoning for this is that installed capacities will continuously be used due to guaranteed feed-in tariffs and transacted investments ("sunk costs"). Thus, the scenarios depict the newly developing bioenergy usages in the upcoming years.

In the context of the scenario analysis, a number of specifications were made which are valid for all scenarios:

- The *composition of cultivation area for the different arable crops* is held unchanged as far as not restricted by crop diversification obligations. Shifts in arable cultivation could not be mapped due to missing information on impacts of the new CAP, imaginable developments of international agricultural prices and/or changes in food demand on the regional production. In the same way, the composition of forests (coniferous, deciduous and mixed forests) is regarded as constant. Reasoning is here the mid-term timeframe of the scenarios and the missing information on possible changes.
- *Agricultural yields* are assumed to develop as in the past (reference period 2000–2010) and are extrapolated at district level for every arable crop (section 4.1.4.2).
- The trend *development of animal production* of the past is not extrapolated. Therewith, no increase of productivity and reduction of the number of animals are assumed until 2030, and the demand for permanent grassland, the straw use for husbandry bedding and the generated manure are not influenced. Reason is that the national assessments on the future development of livestock husbandry cannot be easily transferred to regional and district level.
- No *shift of agricultural and forestry production between the URR and other regions* is foreseen. The current degree of regional supply for the regional demand (of food, wood, bioenergy etc.) is unknown. For example, wood for small combustion plants can origin from regional forestry or national and international wood markets. Additionally, possible changes in competitiveness between regions are uncertain. Therefore, the scenarios indicate only changes in the URR biomass supply, without regarding adjustments with other regions.

- Possible *changes in regional demand for food and wood products* are not part of the scenarios. Such changes would influence (positively or negatively) the potential biomass for energetic usages if the production in the URR is directly influenced. Future changes of demand are very difficult to determine at the regional level.
- *Climate change* can potentially reduce agricultural and forestry productivity. Extreme weather events can disrupt production. This cannot be integrated in the scenarios due to insufficient information about the probable regional effects.

The scenario analysis has a focus on agricultural biomass due to the higher number of different biomass categories with potential relevance for bioenergy production in the URR.

4.1.3 The 'OUI Biomasse' project scenarios

Three scenarios were elaborated and analysed: Scenario 1 "Business as usual (BAU)", Scenario 2 "Maximum Exploitation (MaxEx)" and Scenario 3 "Conservation and Recreation (ConsRec)". Key points of these scenarios are presented in the following subsections.

4.1.3.1 BAU Scenario

In the BAU scenario, current political and economic conditions, which influence the production and use of biomass in the Upper Rhine Region, are kept unchanged. In the field of energy policy, it is assumed that the current regulations and feed-in tariffs stay in force until 2030 (Table 4.1):

- *Renewable electricity*: New regulations of the Renewable Energy Law (Erneuerbare Energie Gesetz – EEG) 2014 (BMWi 2015a) for Germany, current combination of feed-in tariffs for small installations and tendering procedures of projects with higher capacity for France (MESDE 2010), the current cost-reflective feed-in tariffs (Kostendeckende Einspeisevergütung – KEV) and maximum network supplement (Netzzuschlag) for Switzerland (Rieder et al. 2012, BFE 2014) are the key regulations.
- *Biofuels*: In line with the compromise between European Parliament and Council on the brink of adoption, an amendment of renewable fuel target in the Renewable Energy Directive (2009/28/EC) is assumed, restricting conventional biofuels from food crops to 7% of fuel use in transport, which is impacting Germany and France. For Switzerland, a quota system will not be introduced and the fuel tax reduction remains restricted to sustainable biofuels (Schweizerische Bundesrat 2008).
- *Renewable heat*: The current national policies combine the promotion of different renewable energy sources for heating with measures for improving the heating efficiency. This mix of promotion instruments is assumed to be in force until 2030.

This scenario does not represent a trend extrapolation of past developments due to recent changes in the national policies for renewable energy. Additionally, it is assumed that no new bioenergy technologies will become competitive in economic terms until 2030, so that the energetic use of biomass is restricted to proven and already applied technologies (Figure 4.1).

New *biogas* plants based on energy crops will not be built in the BAU scenario. In Germany, biogas production from energy crops is generally no longer profitable with the EEG 2014 feed-in tariffs. In the same way, the

French feed-in tariffs hinder the use of energy crops in biogas production. In Switzerland, the KEV feed-in tariffs allow energy crops only as co-substrate up to 20%. Furthermore, economic reasons speak against the use of energy crops. In Germany, only small-scale biogas plants based on (liquid) manure have restricted changes for expansion (Gömann et al. 2013). The possible technical progress in this biogas category is restricted so that high investment cost degression cannot be expected. Overall, only a slow increase of small-scale manure based biogas plants is assumed, using 20% of calculated manure potential (see section 3.1.2) per district.

Bioenergy production based on agricultural residues (straw) will not become economically feasible in the BAU scenario.

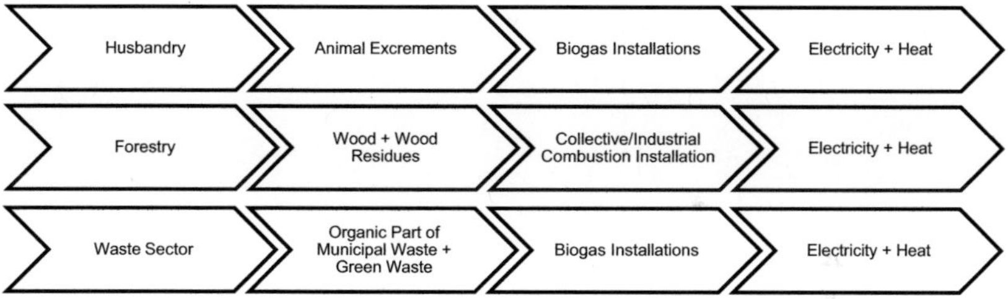

Figure 4.1: Relevant conversion pathways for new bioenergy installations in the BAU scenario.

Table 4.1: Scenario assumptions for the driver "Energy policy".

Policy area	Scenario BAU	Scenario MaxEx	Scenario ConsRec
Renewable electricity			
D: Renewable Energy Act (EEG)	Current feed-in tariffs (EEG 2014) unchanged	Regional tender procedures oriented on the unexploited sustainable biomass	Improved feed-in tariffs for residues + wastes with ecologic restrictions
F: Energy laws (Loi Nouvelle Organisation du Marché de l'Électricité 2010)	Current feed-in tariffs unchanged	Regional tender procedures oriented on the unexploited sustainable biomass	Improved feed-in tariffs for residues + wastes with ecologic restrictions
CH: Energy Law (EnG)	Current feed-in tariffs unchanged	Regional tender procedures oriented on the unexploited sustainable biomass	Improved feed-in tariffs for residues + wastes with ecologic restrictions
Biofuels			
D: EU RES Directive 2009/28/EC + national regulation	7% (resp. national) quota for conventional biofuels	10% (resp. national) biofuel quota	No biofuel quota
F: EU RES Directive 2009/28/EC + national regulation	7% (resp. national) quota for conventional biofuels	10% (resp. national) biofuel quota	No biofuel quota
CH: Fuel tax regulation (MinölStV 2008)	No targets + quota, tax reduction for sustainable biofuels	No targets + quota, tax reduction for sustainable biofuels	No targets + quota, tax reduction for sustainable biofuels

Policy area	Scenario BAU	Scenario MaxEx	Scenario ConsRec
Renewable heat			
D: Renewable heating law 2011 (EEWärmeG) Market Incentive Program	Current promotion	Stronger promotion of CHP and local heating networks	Promotion of CHP and local heating networks, with ecological restrictions
F: Grenelle de l'Environnement	Current promotion	Stronger promotion of CHP and local heating networks	Promotion of CHP and local heating networks, with ecological restrictions
CH: Model cantonal building prescriptions (MuKEn)	Current promotion	Stronger promotion of CHP and local heating networks	Promotion of CHP and local heating networks, with ecological restrictions

Table 4.2: Scenario assumptions for the driver "Agricultural policy".

Policy area	Scenario BAU	Scenario MaxEx	Scenario ConsRec
Ecological focus area			
D, F: CAP 2014-2020 – Greening	5% of arable land	No ecological focus area regulation	10% of arable land, stronger specifications
CH: AP14-17 – Crosscompliance regulation	7% of UAA, extensive production payment for arable land	7% of UAA, extensive production payment for arable land	3% of arable land as fellow land
Crop diversification			
D, F: CAP 2014-2020 – Greening	Current crop diversification requirements – No change in ratio of arable crops	Current crop diversification requirements – No change in ratio of arable crops	Main crop (maize) maximal on 50% of arable land (per district)
CH: AP14-17 – Crosscompliance regulation	Current crop diversification requirements – No change in ratio of arable crops	Current crop diversification requirements – No change in ratio of arable crops	Current crop diversification requirements – No change in ratio of arable crops
Maintenance of permanent grassland			
D, F: CAP 2014-2020 – Greening	Ploughing up forbidden – No change in permanent grassland	Restricted ploughing up allowed – 5% reduction of the permanent grassland in 2010	Ploughing up forbidden – No change in permanent grassland
CH: AP14-17 – Direct payments	No change in permanent grassland	Restricted ploughing up suitable – 5% reduction of the permanent grassland in 2010	No change in permanent grassland

Policy area	Scenario BAU	Scenario MaxEx	Scenario ConsRec
Promotion of organic farming			
D, F: Promotion within 2. Column of CAP	Current promotion system – Conversion trend of the past	Current promotion system – No additional conversions (Status quo)	Improved promotion – 20% share of organic farming on total UAA
CH: AP14-17 – Produktionssystembeiträge Bio	Current promotion system – Conversion trend of the past	Current promotion system – No additional conversions (Status quo)	Improved promotion – 20% share of organic farming on total UAA

Renewable electricity production from solid biomass is mainly based on the *combustion of woody biomass*. The waste wood, saw mill and industry residue wood potentials were mobilised in the first years after the EEG 2000. Beginning with the bonus system of the EEG 2004, also forest residue wood is used. Energy wood prices increased in the last years and led to lower production capacity of new installations. The result was a continuing growth of the number of combustion units, but with increasingly slower growth of electricity production capacity and production (BMWi 2015b, p. 4). With the EEG 2014 feed-in tariffs, new exploitation of energy wood (forestry residue wood) potentials is becoming very difficult. The same applies for the French and Swiss part of the URR because the current feed-in tariffs are on a similar level. The French renewable energy policy gives some priority to renewable heat production and the support of biomass plants for renewable electricity is restricted to 5 to 12 MW. In addition, the already relative high level of energy wood use in the region makes the mobilization of wood from the remaining potentials more expensive and difficult. Based on these considerations, it is assumed that in the BAU scenario only 10% of the calculated energy wood potential will be used additionally in 2030.

In the *heating sector*, the number of modern biomass heating systems (pellet, woodchip and split logs heaters) will increase. Partly, these new more efficient installations will replace old biomass boilers and stoves. As far as an additional energy wood demand is triggered, it is assumed that the supply comes from national and international energy wood markets. Some additional heat supply comes from new combined heat and power (CHP) plants (s.o.).

The parameters of agricultural policies (Table 4.2) regarding Germany and France are based on the current reform of the Common Agricultural Policy (CAP 2014-2020). Consequently, guidelines stemming from the new "Greening" architecture of the CAP (ecological focus areas, crop diversification, and maintenance of permanent grassland) can influence the supply of agricultural biomass usable for the bioenergy production. For Switzerland it is presumed that current agricultural policies continue to be valid. In all three countries, conversion to organic farming will continue according to trends of the past. Policies with regard to nature protection and environmental policies will not be changed; consequently no new limitations for the energetic use of biomass will emerge.

4.1.3.2 MaxEx Scenario

In the MaxEx scenario, it is assumed that the general political and economic conditions will favour an expansion of the energetic biomass usage until 2030. Altogether, a change of conditions is presumed in a way that allows using major parts of the regional biomass potential.

In the *energy policy* context, it is recognised that, in order to reach climate protection goals, the usage of bioenergy has to be increased considerably for the production of *renewable electricity*. Due to the ability of bioenergy to contribute to basic load energy provision, bioenergy can be used to balance other, fluctuating renewable energies. Key instrument for the promotion of renewable electricity will be tender procedures in the MaxEx scenario (Table 4.1). Tender procedures are already used in France for biomass installations with higher capacity. Switzerland is planning to introduce tender procedures for specific plant types and technologies (BFE 2014a, p. 82). The German EEG 2014 regulates that from 2017 on the financial support for renewable electricity should be determined by tender procedures. The EEG lets open the concrete design of calls for tender which is currently in the political process (Kahles 2014). Precondition for the extended biomass usage is the willingness to increase the supplements (EEG-Umlage, KEV-Umlage, etc.) which is assumed. This is necessary to finance the foreseen biomass exploitation for bioenergy.

For the first major conversion pathway *biogas*, great part of the calculated technically available manure (manure of livestock units with more 50 GVE cattle, 100 GVE swine and 100 GVE poultry, see section 3.1.2) is mobilised. The assumed mobilisation rate of 80% represents that biogas production is not economically valid in every case even if the number of livestock is sufficient. Assumption relating to energy crops is that a limited additional cultivation of *energy crops* is possible in the French and German sub-regions due to yield increases and the restricted ploughing up of permanent grassland in this scenario. Deviating, no energy crop cultivation is assumed for Northwest Switzerland due to the political goal to maintain the current domestic level of food production and the critical perception of energy crops in Switzerland. Additionally, it is assumed that the new energy crop area in 2030 is used in equal shares by crops for biogas production and lignocellulose crops for combustion or gasification. The second major conversion pathway is combustion or gasification of lignocellulose biomass. It is assumed that 80% of calculated *crop residue* potential (see section 3.1.2) is mobilised. As a third biomass resource for combustion or gasification, 80% of the calculated energy wood potential (*forestry residue wood*) will be mobilised in the MaxEx scenario. Therewith, besides already used biomass categories and applied conversion pathways, new pathways and until now unused biomass categories are introduced in bioenergy production (Figure 4.2).

It is assumed that original European *biofuel* targets (10% of fuel consumption in 2020) of the Renewable Energy Directive 2009/28/EC are maintained and prolonged until 2030. This will lead to increasing biofuel consumption within the EU, which, consequently, requires more biomass for the production of biofuels. It is assumed that no additional agricultural area for energy crops is needed in the EU to achieve the 10% goal in 2030 due to imports and introducing 2nd generation biofuels[2]. In the *renewable heat* sector, the promotion activities are redirected from biomass heating systems of single buildings to CHP plants, often in combination with local heating networks. Aims are to achieve higher efficiency and reduce emissions from biomass heating. The new alignment opens also the use of tender procedures, together with such for renewable electricity. The biomass source for the additional renewable heat and electricity is intended to be as follows: In simplification, the increasing biomass heat production is solely allocated to CHP plants and based on the forestry residue wood consumption for electricity production. In consequence, it is assumed that no addi-

[2] 1st generation biofuels are produced from food-crops (cereals, sugar crops, oil crops), 2nd generation biofuels are based on non-food-crops, lignocellulose plants and organic residues and wastes.

tional demand for forestry residue wood is caused by the development in the heating sector. This is in line with the expectation that the final energy consumption of biomass for heating might even decline, especially after 2020, in combination with an increasing share in the heating market (Kranzl et al. 2013).

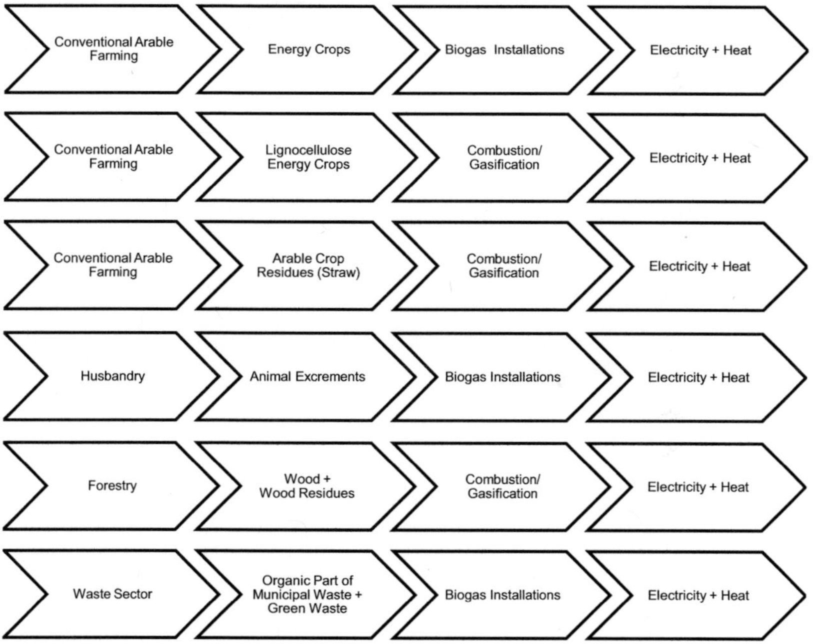

Figure 4.2: Relevant conversion pathways for new bioenergy installations in the MaxEx scenario.

General conditions regarding *agricultural and environmental policies*, as well as nature protection, basically stay the same as in the BAU scenario (Table 4.2). However, since the focus lies on increased bioenergy production, "Greening" requirements (of the EU Common Agricultural Policy (CAP)) are withdrawn. In the next CAP period starting from 2020, ecological focus areas are no longer demanded. As in the CAP before 2014, a restricted ploughing up of permanent grassland is again allowed. Full use is made of the permissible amount of ploughing up of grassland (5%), caused by economic feasibility of bioenergy usage.

In this scenario, the competition for leased land by farmers cultivating energy crops will be high. This will restrict the expansion of organic farming. As seen in the past, biogas production in Germany was very competitive in economic terms under the payment conditions of the Renewable Energy Act (EEG) 2009 (Meyer, Priefer 2012), influencing also the development of organic farming. Additionally, high demand for biomass from bioenergy production will probably stabilise or increase the prices for farm products. High agricultural prices reduce the incentives for a conversion to organic farming. Based on these considerations, it is assumed for the scenario MaxEx that no more conversion to organic farming will take place until 2030. Therefore, the utilised agriculture area used by organic agriculture from 2010 remains unchanged.

4.1.3.3 ConsRec scenario

This scenario represents a development in which ecological requirements will grow in importance. Nature protection and the preservation of landscape play a more important role which is represented within the agricultural policy assumptions. This impacts the way of agricultural land use. The ecological orientation of the scenario influences both, the bioenergy policies and the basic conditions for using biomass. It is predetermined that biomass exploitation has to be done in a resource-saving way, which means that only parts of the biomass available will be used. Energetic use of biomass is restricted to residues and waste (Figure 4.3).

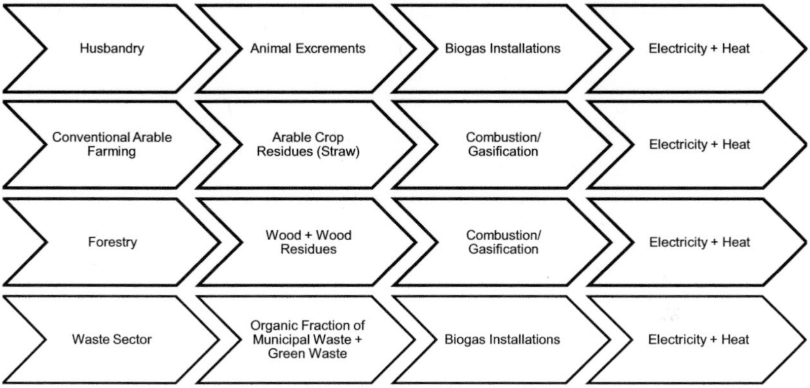

Figure 4.3: Relevant conversion pathways for new bioenergy installations in the ConsRec scenario.

For *renewable electricity*, the promotion for the use of agricultural and forestry residues is improved in the ConsRec scenario, in contrast to the current conditions (BAU scenario). This is achieved by higher feed-in tariffs and/or tender procedures for renewable electricity from agricultural residues and manures and from forestry residue wood (Table 4.1). This requires an amendment of the German EEG, the Swiss KEV regulation and the French procedures.

For *biogas* production, it is assumed that 80% of the calculated technically available manure is used. The high mobilisation rate (by appropriate incentives) is regarded as adequate because biogas production from manure is not associated with new negative environmental impacts in comparison to direct manure application. The use of *agricultural residues* (especially straw) for combustion is also promoted. The assumption is that 80% of calculated residue potential can be mobilised. The residue potential in the ConsRec scenario is lower than in the MaxEx scenario because stronger requirements for the maintaining of soil fertility are applied in the calculation of residues for the ConsRec scenario. Third biomass resource used in the MaxEx scenario is *forestry residue wood*. It is assumed that 50% of the calculated energy wood potential can and will be used. This assumption (with the lower percentage as in the MaxEx scenario) reflects that more residue wood should remain in the forest for nutrient recycling and soil fertility maintenance, reflecting the ecologic orientation of this scenario.

The stronger ecological orientation of this scenario is also reflected in future *biofuel* policies. The critics on the current biofuel regulation, especially of environmental NGOs, include indirect land use changes destroying natural forests (e.g., rainforests), insufficient greenhouse gas emission reduction or even increasing emissions due to land use change, induced loss of biodiversity and negative impacts on smallholders and sustainable farming. In the ConsRec scenario, these arguments against biofuels will be successful in the future. Assumed changes in the biofuel policy are no biofuel quota and no 1st generation biofuel from food/feed crops. In consequence, the virtual land use of biofuel crops in the URR will be reduced to zero in 2030.

Renewable heat promotion takes a similar approach as in the MaxEx scenario. The increase in biomass heating is restricted by the stronger ecological obligations in regard to the exploitation of forestry residue wood. The additional use of forestry residue wood is once again allocated to renewable electricity, assuming combined production in CHP plants.

In the ConsRec scenario, it is assumed that environmental objectives of *agricultural policy* will become more prominent in the future and direct payments will become more targeted to the provision of public goods (Table 4.2). For the post-2020 CAP of the EU, this is represented by revised and stronger Greening requirements:

- Ecological focus areas will be extended to 10% of the area of arable land.
- Requirements regarding crop diversification are tightened (e.g., cultivation of the main crop limited to 50% of the area of arable land).
- Ploughing up of permanent grassland is completely prohibited.

Organic farming is regarded as having to play a pioneering role in sustainable agriculture. In this context, a change of the EU CAP is part of the scenario. In the result, economic barriers to conversion are removed because the relative economic efficiency of organic farming over conventional agriculture will be strengthened by higher production costs for conventional agriculture and significantly improved support of organic farming under the second pillar of the CAP. Background is additionally that the demand for organic food increases stronger than in the other scenarios. The German sustainability strategy formulates as a goal a 20% share of the total utilised agricultural area (UAA) for agricultural land used by organic farms. This goal is taken to simulate the development of organic farming in the scenario ConsRec. It is assumed that in the French, German and Swiss part of the URR respectively a share of 20% will be reached overall.

4.1.4 Common developments in the scenarios

Beside the scenario specific assumptions, agricultural land use and production is influenced by other factors, which are not influenced by and differentiated between the scenarios. The future availability of agricultural land is influenced by land use change (section 4.1.4.1): the transformation of agricultural land into artificial areas for housing, industry and infrastructure. The future agricultural production is dependent on the development of yields (section 4.1.4.2).

4.1.4.1 Land use change

Based on the development from 2000 to 2010, the future land use change is assessed (Table 4.3). The relation of settlement and transport area increase to agricultural land use change in the period 2000 – 2010 is used to determine the land use change caused by extension of settlement and transport areas. Therefore, the additional assumption is that in the next two decades the structural changes in agriculture will not lead to abandonment of utilised agricultural area (UAA) due to relatively high agricultural product prices and high international demand.

It is assumed that the slowdown trend of increasing settlement and transport area during the last decade (Destatis 2014, p. 15) will continue in the future and the sustainability goal of a further reduction of land consumption (e.g., SL BW 2013) will be achieved. Therefore, the rate of loss of utilised agricultural area per year in the period 2010 – 2030 is reduced to half of the rate in the period 2000 – 2010.

No change of UAA is assumed for the districts where in the reference period an increase of UAA has taken place. This is based on the fact that no real land reserves are available for a further extension of UAA. The development of UAA is assigned solely to reduction of arable land. Changes between permanent grassland and arable land are determined by agricultural policy regulations and are scenario dependent.

Table 4.3: Land use change in the period 2010 – 2030.

Sub-region and district	Change of UAA in the period 2000 – 2010 [1)] (%)	Relation of settlement and transport area increase to agricultural land use change in the period 2000 – 2010 (%)[2)]	Assumed change rate of UAA in the period 2010 – 2030 (%)
Rheinland-Pfalz	0.2	-	-
Landau in der Pfalz	-9.6	24	- 2.5
Germersheim	3.4	- [3)]	0
Südliche Weinstraße	-1.1	145	- 1.0
Baden-Württemberg	-6.3	-	-
Baden-Baden	-6.9	33	- 2.3
Freiburg	-4.3	108	- 4.0
Karlsruhe Stadt	-4.3	239	- 4.0
Karlsruhe Land	-0.4	1 020	- 0.5
Rastatt	-0.2	1 350	- 0.2
Breisgau-Hochschwarzwald	-6.8	18	- 1.4
Emmendingen	-7.0	32	- 2.3
Ortenaukreis	-9.3	26	- 2.3
Lörrach	-8.0	34	- 2.6
Waldshut	-7.3	20	- 1.4
Alsace	0.1	-	-
Bas-Rhin	0.5	-	0
Haut-Rhin	-0.4	-	0
Northwest Switzerland	-0.4	-	-
Aargau	-2.1	158	- 2.0
Basel-Stadt	-16.3	15	- 2.5
Basel-Landschaft	-1.5	177	- 1.5
Solothurn	-1.0	299	- 1.0
Jura	3.5	- [3)]	0

Notes: 1) Period 2003 – 2010 for UAA change in Rheinland-Pfalz and period 1999 –2010 for UAA change in Baden-Württemberg

2) Total increase in settlement and transport area 2000 – 2010 / Total decrease in UAA 2000 – 2010

3) Relation not applicable due to increase in utilised agricultural area

4.1.4.2 Yield development

Future yield developments are extrapolated from the yield trend in the period 2000–2010 for every crop and district (see section 3.1.2). The relative yield increases from 2010 to 2030 show a high variety, ranging from unchanged up to 52% increase, and is depending on the specific arable crop and the location (Table 4.4).

Table 4.4: Range of yield increase 2010 – 2030.

Crop	Yield increase 2010 – 2030 (%)
Cereals	
Wheat	2 – 19
Rye	0 – 23
Barley	0 – 20
Oat	0 – 23
Maize	5 – 20
Other cereals	0 – 23
Forage crops	
Silage maize	0 – 9
Cereals as total crops	0 – 9
Legumes	0
Temporary grassland	0 – 16
Root crops	
Potatoes	0 – 32
Sugar beet	9 – 20
Oil crops	
Rapeseed	0 – 52
Sunflower	17 – 19
Soy	16 – 31
Legumes	0 – 13

4.1.5 Agricultural land use and biomass production in the scenarios

This section presents main results of the scenario analysis.

4.1.5.1 Agricultural land use

The total *utilised agricultural area* (UAA) decreases in the scenarios only to a small amount due to land use change (Table 4.5). Only in the MaxEx scenario, the arable land increases due to a restricted ploughing up of permanent grassland.

Table 4.5: Utilised agricultural area (UAA) and main land use categories.

Main land use categories	State 2010 (ha)	BAU 2030 (ha)	MaxEx 2030 (ha)	ConsRec 2030 (ha)
Arable land	466 702	460 430	473 454	460 430
Permanent grassland	260 497	260 497	247 472	260 497
Permanent cultures	52 730	52 730	52 730	52 730
UAA total	782 826	776 554	776 554	776 554

Note: The difference between UAA total and the sum of the main land use categories is based on discrepancies in the statistical data base.

Stronger changes take place between *conventional and organic farming*. The share of organic farming on the total UAA is shown in Figure 4.4. Depending on the agricultural policies in the scenarios, shifts in the area of conventional and organic agricultural land will take place until 2030 as depicted in Figure 4.5. In the BAU and ConsRec scenarios, the conventional UAA is reduced due to conversion to organic farming. In the MaxEx scenario, the conventional arable land increases due to restricted ploughing up of permanent grassland, no new conversion to organic farming and no obligation of ecological focus areas. This improves the chances for more energy crop cultivation and energetic use of crop residues.

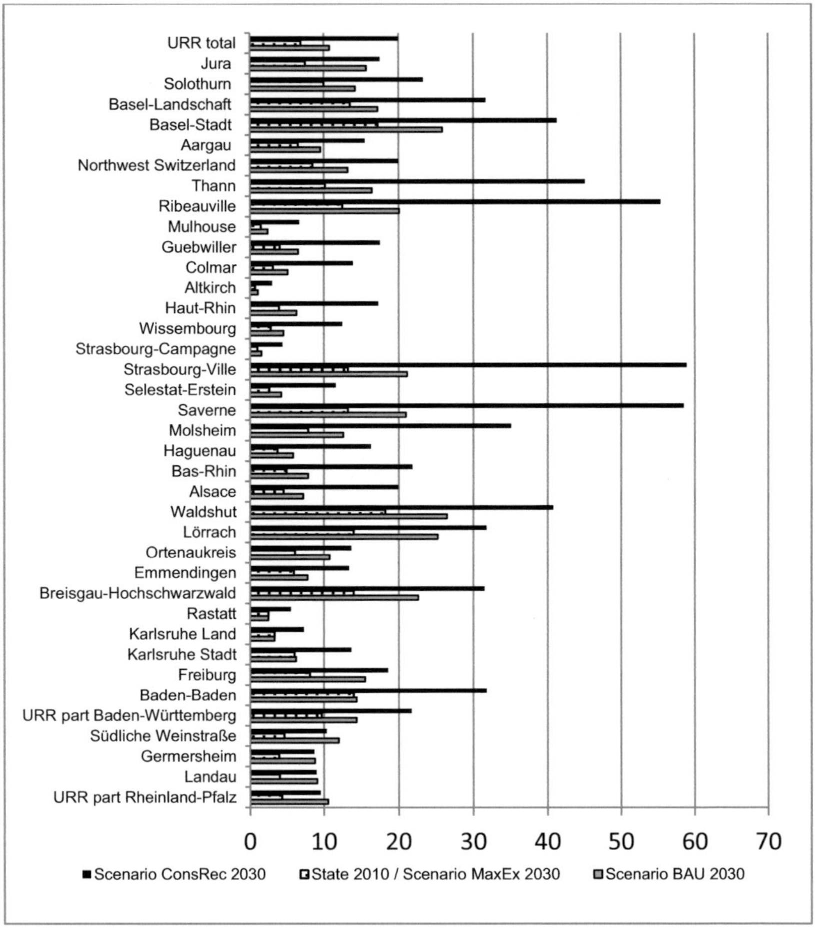

Figure 4.4: Share of organic farming on the overall UAA in the scenarios.

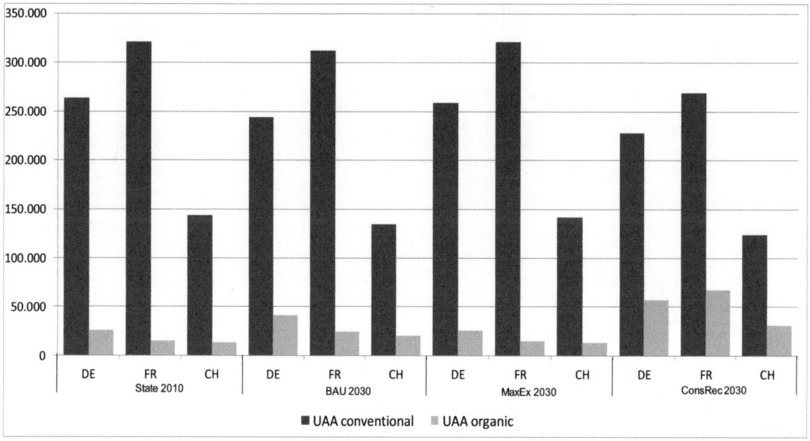

Figure 4.5: Development of conventional and organic UAA (in ha).

Potential production of feedstock for bioenergy is distinctively different in conventional and organic agriculture. In organic farming, cultivation of energy crops is in general economically not viable due to the premium prices for organically produced food. Regarding agricultural residues and wastes, the aim of organic farming to recycle nutrients and improve soil fertility excludes the use of crop production residues (e.g. straw) for energy production.

Beside the development of organic farming, biomass production for energetic use is restricted by scenario dependent *ecological focus area* obligations. The changes in arable land use is shown in Figure 4.6. In the BAU scenario, requirements regarding the ecological focus area are already fulfilled by current land use (legumes area, fallow land and cultivation of intertillage crops at district level) so that no change in the arable cropping is needed. Although the ecological focus area obligations are fulfilled with the continuing of the current cropping scheme at district level, some individual farms might have to take measures to achieve their 5 percent of ecological focus area.

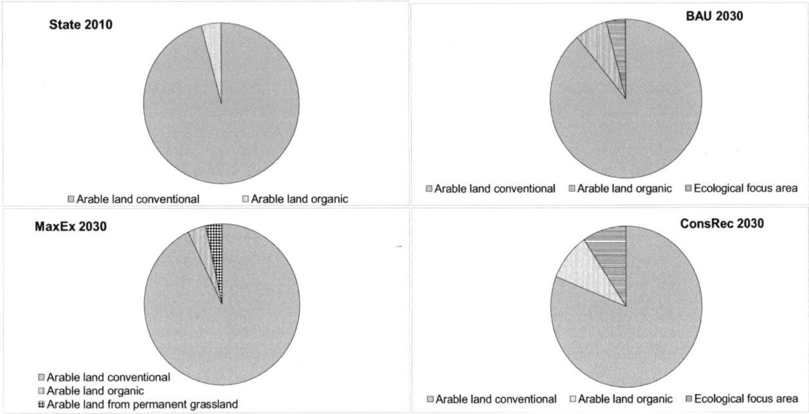

Figure 4.6: Arable land use change in the scenarios.

The *arable land use* is distinctive different between sub-regions as well as between districts. Cereals and especially corn maize are the most important arable crops in the URR. This great variety is of relevance for the available crop residues (see section 4.1.5). Stronger crop diversification as demanded by the current CAP Greening regulations will affect the share of corn maize, especially in same districts of Alsace, as seen in Figure 4.7.

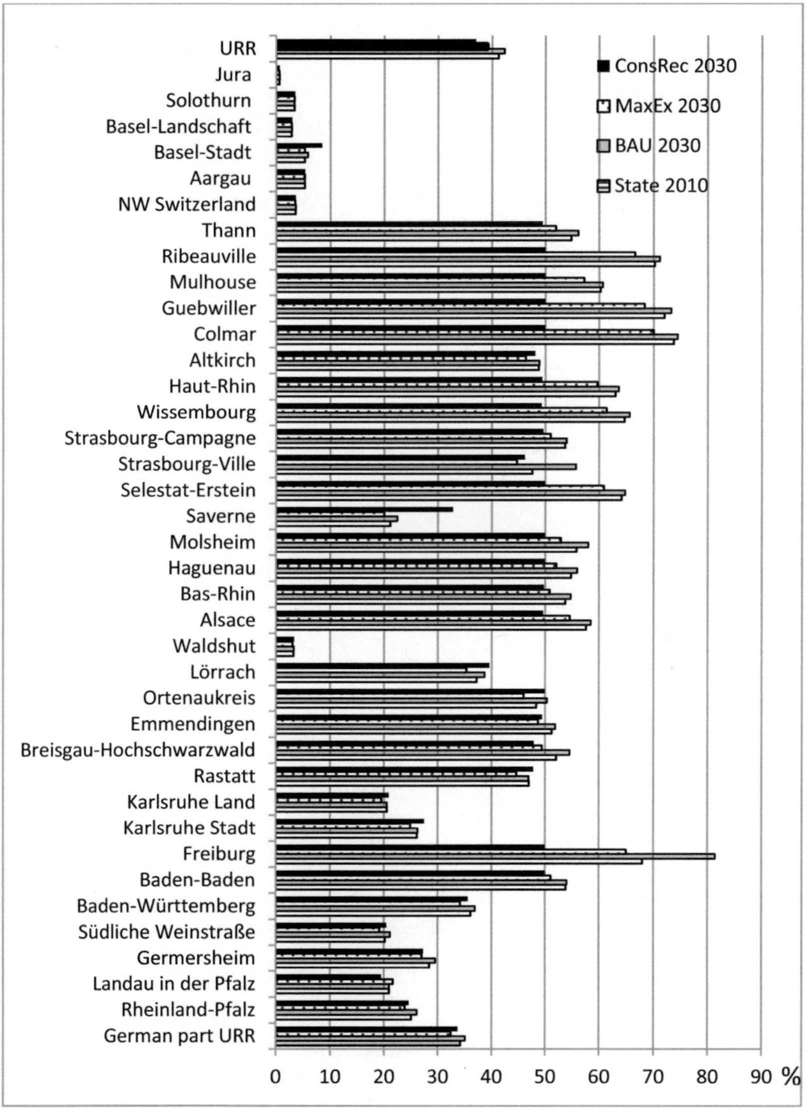

Figure 4.7: Share of corn maize on total conventional arable land in the scenarios (in %).

Energy crops currently occupy in the German part of the URR 8.9% of the arable land and in the French part 2.6%. The "virtual land use" for biofuels remains unchanged in the BAU and MaxEx scenario with 6 033 ha in France and 5 266 ha in Germany (Figure 4.8). Beside the continuing cultivation of energy crops for the existing biogas plants in Germany, energy crop cultivation for regional bioenergy production increases in the MaxEx scenario so that overall energy crops have in this scenario a share of 13.9% in the German URR and 5.0% in the French URR. Therewith, the energy crop percentage in the MaxEx scenario for the German part of the URR remains lower than the current (year 2013) average share of energy crops in total Germany with 16.7% (BMEL 2014, FNR 2015). The ConsRec scenario describes a situation without 1st generation biofuel use in the EU. In Switzerland, energy crops will not be cultivated at all.

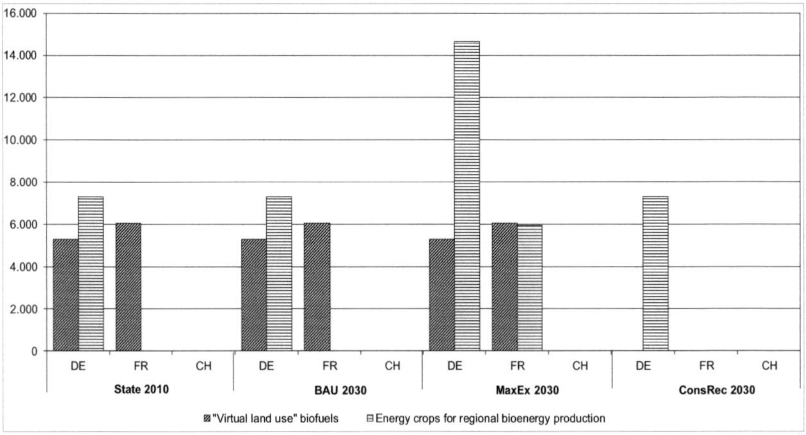

Figure 4.8: Energy crop areas in the scenarios (in ha).

4.1.5.2 Agricultural biomass production

In comparison to the reference situation 2010, the total *production for food and feed* is higher in all scenarios (Figure 4.9), even in the MaxEx scenario with additional energy crop cultivation. But with the stronger ecological focus, the production increase is smaller in the ConsRec scenario. Exemption is Northwest Switzerland, where the food and feed production slightly decrease in the BAU and ConsRec scenario.

Organic food production doubles in the BAU scenario and more than triples in the ConsRec scenario in 2030, compared to reference year 2010. Conventional food and feed production of overall URR increases at lower rates of around 18% in the BAU scenario, 27% in the MaxEx scenario and 13% in the ConsRec scenario.

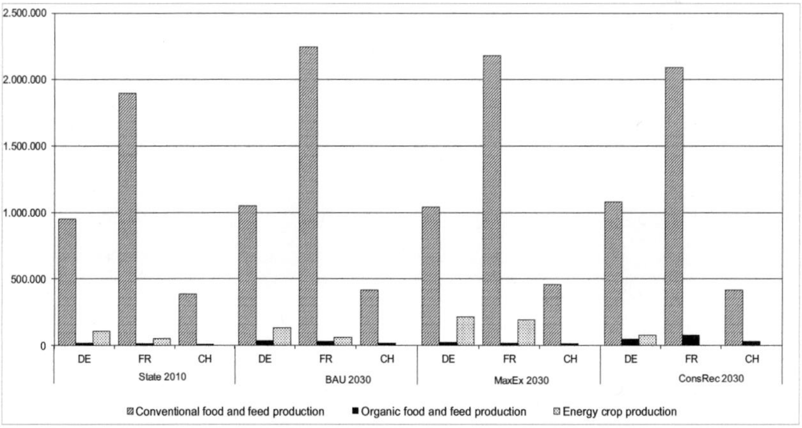

Figure 4.9: Biomass production in the scenarios (in t dry matter).

All three scenarios show a restricted role for *energy crop production* in the URR. Energy crop production in the URR will account for 2.2% (ConsRec scenario) to 6.5% (MaxEx scenario) of the total biomass production in 2030. The increasing food and feed production in the scenarios indicates that there could be potential to cultivate more energy crops in France and Germany than assumed in the scenarios, if the yield trend of the past will continue and higher yields are not associated with increased negative environmental impacts.

The biomass from *arable crop cultivation residues* for bioenergy production differs considerably between the URR sub-regions and the scenarios

(Figure 4.10). Under current conditions (BAU scenario), a mobilisation of crop residues will not take place. Stronger ecological standards (ConsRec scenario) will limit the energetic use of crop residues.

Potential contribution of cereal residues (straw) is overall low and in some districts even not existing (Figure 4.11). Crop residues for bioenergy production in Alsace are strongly dependent on the partial use of corn maize residues. Overall, crop residues for bioenergy production would strongly be reduced without residues from corn maize. Residues from oil crop and sugar beet cultivation are marginal (around 10 000 t dry matter per sub-region and crop category).

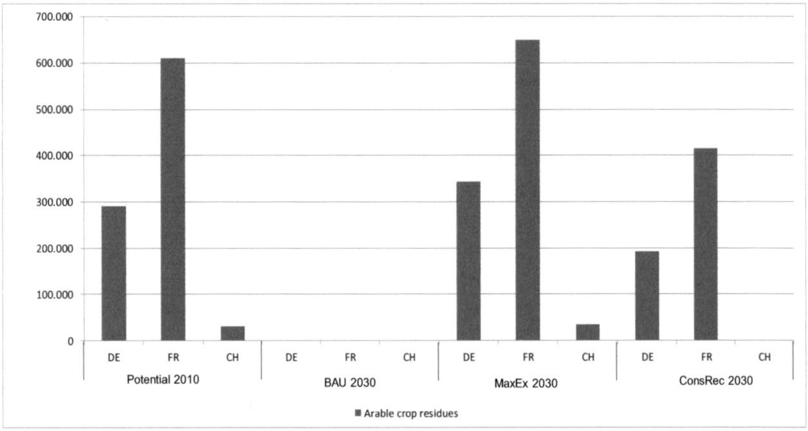

Figure 4.10: Arable crop residues for bioenergy production in the scenarios (in t dry matter).

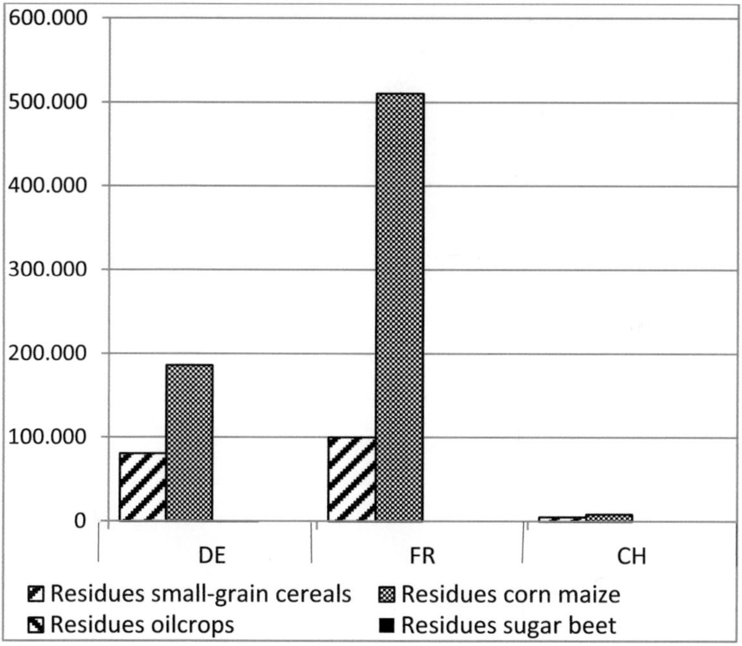

Figure 4.11: Composition of arable crop residues for bioenergy production (Potential 2010 in t dry matter).

The biomass potentials from *manure* (in dry matter) are lower than the residue potentials. The highest potential for biogas production from husbandry manure is seen in Northwestern Switzerland due to high husbandry density (Table 4.6). Under current conditions (BAU scenario), the mobilisation of manure for biogas is restricted. Higher amounts of manure are available in the MaxEx and ConsRec scenarios due to the assumed mobilisation rate of 80% of the calculated manure potential. Differences between the sub-regions in regard to livestock composition and number of livestock per farm influence the available manure for biogas production. Liquid manure from cattle husbandry are dominating the available manure.

Table 4.6: Manure for biogas production in the scenarios.

Sub-region / Scenario	Husbandry density[1)] (LSU/ha)	Manure density[1)] (Total manure per UAA in kg/ha)	Total available manure for biogas production[2)] (t/y)
State / Potential 2010			
Germany	0.48	311	43 833
France	0.28	211	40 486
Switzerland	1.35	1 057	77 658
BAU scenario 2030			
Germany	0.49	316	8 767
France	0.28	211	8 097
Switzerland	1.37	1 070	15 532
MaxEx scenario 2030			
Germany	0.49	316	35 066
France	0.28	211	32 389
Switzerland	1.37	1 070	62 127
ConsRec scenario 2030			
Germany	0.49	316	35 066
France	0.28	211	32 389
Switzerland	1.37	1 070	62 127

Notes: 1) Husbandry includes cattle, swine, sheep, goat, horses, without poultry
2) Calculation of manure for biogas production is based only on manure from cattle and swine, and includes only farms with over 50 LSU cattle or 100 LSU swine.

4.1.6 Forestry biomass in the scenarios

Based on the forest inventories, potential for additionally available wood residues from wood harvest were calculated (see section 3.1.2). In the BAU scenario with the current conditions of bioenergy promotion, only a small part of wood residues will be mobilised until 2030 (Figure 4.12). Higher incentives for bioenergy (MaxEx scenario) will lead to a distinct increase of forestry wood residue use. But in relation to the already used energy wood, the amount of additional wood residue for bioenergy production will remain restricted. Stronger criteria for recycling biomass in the forest and an extension of nature protection areas (ConsRec scenario) will restrict the use of wood residues for bioenergy production. Additionally, the expanding certification of forests (e.g., the FSC[3] certification of state forest in Baden-Württemberg) limits the extraction of residue wood. Overall, energy wood harvesting and use is already high in the URR so that additional use is restricted and in some districts not possible while respecting ecological restrictions.

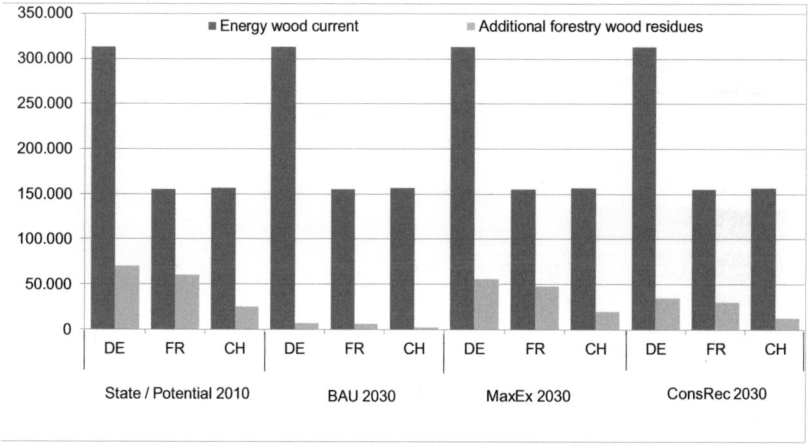

Figure 4.12: Energetic used biomass from forestry (in t wood per year).

4.1.7 Waste biomass in the scenarios

Improved use of the biogenic fraction in municipal waste for bioenergy production can be achieved by different pathways:

- Separate collection of bio-waste ("Biotonne") with following fermentation (biogas production) before composting
- Redirection of bio-waste from direct composting to fermentation (biogas production) with subsequent composting

[3] Forest Stewardship Council

- Redirecting part of the green waste (public collection of private garden residues, residues from public parks, roadside vegetation etc.) from direct composting to fermentation (biogas production) before composting
- Improved collection of green waste with following fermentation (biogas production) before composting

Under current conditions (BAU scenario), only part of the bioenergy potential from municipal biogenic waste fraction can be redirected from incineration to fermentation due to barriers such as delays in the build-up and improvement of separate collection systems, the transacted investments ("sunk costs") in composting installations and waste incineration plants, the low feed-in tariffs (respective the phase-out of feed-in tariffs in Switzerland), and local conflicts around waste fees. In the MaxEx scenario, one focus is on the bioenergy production from household wastes to the extent that the potentials are more or less used. With the ecological orientation of the ConsRec scenario, the focus is on the use of residues and wastes which is considered more environmental-friendly, but the quantity of bio-waste is somewhat reduced due to a strong policy to combat food waste.

4.1.8 Conclusions for section 4.1

Agricultural land use in the URR is today dominated by food production which is maintained in all scenarios. In forestry, the production of energy wood plays already an important role. The additional biomass for bioenergy production, as worked out in the scenarios, remains restricted. Figure Figure 4.13 shows the energy content of the analysed biomass categories from agriculture and forestry for energetic uses. The contribution of these energy contents to final energy supply is depending on the efficiency of the conversion pathways.

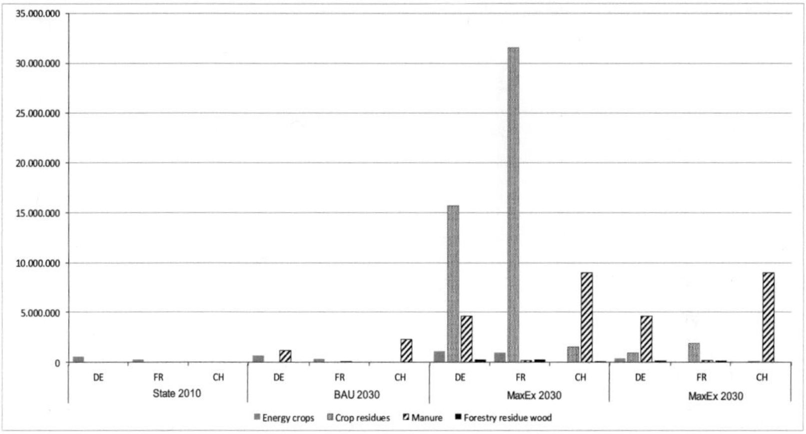

A) Energy content of biomass from energy crops, crop residues, manure and forestry residue wood, state 2010 and scenarios BAU, MaxEx and ConsRec in 2030

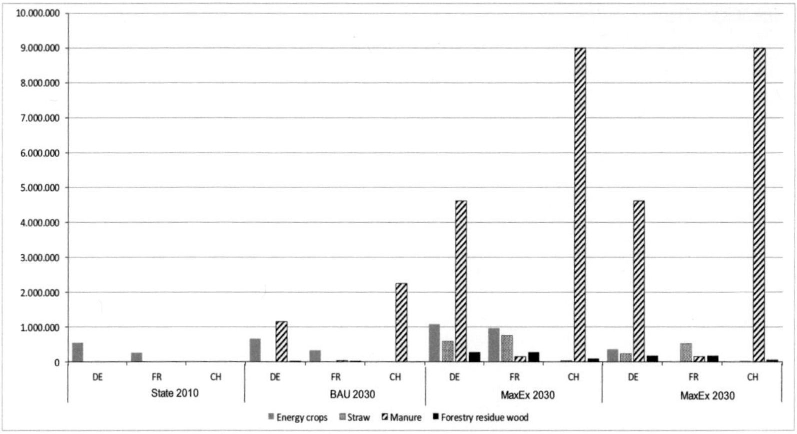

B) Energy content of biomass from energy crops, only cereal straw (without residues from corn maize), manure and forestry residue wood, state 2010 and scenarios BAU, MaxEx and ConsRec in 2030

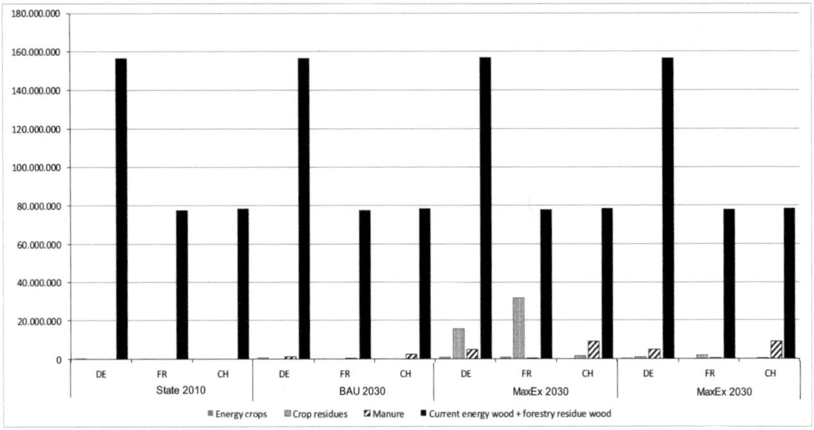

C) Energy content of biomass from energy crops, crop residues, manure, forestry residue wood and current energy wood use, state 2010 and scenarios BAU, MaxEx and ConsRec in 2030

Figure 4.13: Energetic content of biomass used for bioenergy production (in MWh).

Regarding the energy content of the analysed biomass categories, the contribution of energy crops and forestry residue wood is low. In the MaxEx scenario, a substantial contribution comes from crop residues in the German and French sub-region (Figure 4.13 A). But this is only the case when residues from corn maize cultivation can be mobilised in an ecologically agreeable way. With the stronger ecological restrictions in the ConsRec scenarios, the contribution of crop residues is reduced considerably. For straw from cereals without residues from corn maize, the energy content is even lower than the energy content of biomass from energy crop cultivation (Figure 4.13 B). The second major additional biomass category is manure for biogas production (Figure 4.13 A and B). The contribution of manure, in terms of energy content, is of relevance in the German and Swiss part of the URR, but insignificant in Alsace. In comparison to the current energy wood production, the mobilisation of additional forestry residue wood is marginal (Figure 4.13 C).

In all scenarios, the already established energetic use of wood from forestry remains predominant. The contribution of all included crop residues achieve only around 40% (MaxEx scenario France) of the energy content of the current energy wood use respectively manure around 11% (ConsRec scenario Switzerland). Overall, crop residues and manure are the most promising biomass categories for a sustainable extension of bioenergy production in the URR, but also associated with remarkable uncertainties and obstacles.

4.1.9 References section 4.1

BFE (Bundesamt für Energie) 2014a, *Wirkung der Systeme zur Förderung von Elektrizität aus erneuerbaren Energien*. Bericht in Erfüllung des Postulates 09.3085 (Guy Parmelin) vom 12. März 2009. Vom Bundesrat an seiner Sitzung vom 20.06.2014 gutgeheissen.

BMEL (Bundesministerium für Ernährung und Landwirtschaft) 2014, *Statistisches Jahrbuch über Ernährung, Landwirtschaft und Forsten 2014. Landwirtschaftlich genutzte Fläche nach Kulturarten*, Available from: <http://berichte.bmel-statistik.de/SJT-3070400-0000.pdf>.

BMWi (Bundesministerium für Wirtschaft und Energie) 2015a, *Erneuerbare-Energien-Gesetz 2014,* Available from: <http://www.bmwi.de/DE/Themen/Energie/Erneuerbare-Energien/eeg-2014.html>.

BMWi (Bundesministerium für Wirtschaft und Energie) 2015b, *Marktanalyse Biomasse,* Available from: <http://www.bmwi.de/BMWi/Redaktion/PDF/M-O/marktanalyse-freiflaeche-photovoltaik-biomasse,property=pdf,bereich=bmwi2012,sprache=de,rwb=true.pdf>.

Destatis (Federal Statistical Office of Germany) 2014, *Sustainable Development in Germany. Indicator Report 2014*, Available from: <https://www.destatis.de/EN/Publications/Specialized/EnvironmentalEconomicAccounting/Indicators2014.pdf?__blob=publicationFile>.

FNR (Fachagentur Nachwachsende Rohstoffe) 2015, *Anbau Nachwachsender Rohstoffe in Deutschland für die Jahre 2013/14,* Available from: <http://mediathek.fnr.de/media/downloadable/files/samples/a/b/abb03_rgb.jpg>.

Gömann, H, de Witte, T, Peter, G, Tietz, A 2013, *Auswirkungen der Biogaserzeugung auf die Landwirtschaft*, Thünen Report 10, Johann Heinrich von Thünen-Institut, Braunschweig.

Knapp, M, Schumacher, K, Weinberger, N 2014, *Scenarios for potential biomass futures in the tri-national upper rhine region*. In: Michalek, T, Hebáková, L, Hennen, L, Scherz, C, Nierling, L, Hahn, J (eds.): Technology assessment and policy areas of great transitions. Proceedings from the PACITA 2013 conference in Prague. Prague, Czech Republic: Informatorium 2014, pp. 237-242.

Kranzl, L, Hummel, M, Müller, A, Steinbach, J 2013, Renewable heating: Perspectives and the impact of policy instruments. *Energy Policy*, vol. 59, pp.44-58.

MESDE (Ministry of Ecology, Sustainable Development and Energy) 2010, *Énergie de biomasse et biogaz*. 1er septembre 2010, Available from: <http://www.developpement-durable.gouv.fr/-Energie-de-biomasse-et-biogaz-.html>.

Meyer, R, Priefer, C 2012, *Ökologischer Landbau und Bioenergieerzeugung – Zielkonflikte und Lösungsansätze*. TAB-Arbeitsbericht Nr. 151. Büro für Technikfolgen-Abschätzung beim Deutschen Bundestag, Berlin.

Rieder, S, Bernath, K, Walker, D 2012, *Evaluation der kostendeckenden Einspeisevergütung (KEV)*, Luzern, Zollikon, Interface Politikstudien Forschung Beratung, Ernst Basler + Partner, Université de Genève.

Schweizerische Bundesrat 2008, *Mineralölsteuervorordnung (MinöStV)*. Änderung vom 30. Januar 2008, Available from: <http://www.news.admin.ch/NSBSubscriber/message/attachments/10941.pdf>.

SL BW (Statistisches Landesamt Baden-Württemberg) 2013, *Täglicher Flächenverbrauch*, Available from: <https://www.statistik-bw.de/BevoelkGebiet/Indikatoren/GB_flaechenverbrauch.asp>.

4.2 Identification of potential production sites

***Andreas Rudi**, Ann-Kathrin Müller and Frank Schultmann*

French-German Institute for Environmental Research (DFIU), Karlsruhe Institute of Technology (KIT), Karlsruhe, Germany.

The contribution of this research area is manifold. Firstly, based on the provided data of research areas (RA) 1 (see section 3.1), and the scenario results of RA 4 (see section 4.1), as well as RA 3 (see section 3.2) the objective of RA 2 (biomass value chain and logistics) is to link biomass potentials with conversion technologies for energetic and material utilisation. Secondly, potential locations of the conversion facilities are identified and their capacities determined. However, given the areas of biomass provision, for simplicity reasons defined as central locations (sources), and potential conversion plant locations (sinks), the planning of biomass distribution, the choice of the most suited conversion pathway with the most economic capacity is not trivial and requires decision support.

The design of biomass value chains depends on the trade-off between transportation costs and economies of scale of technology capacity investments. Due to the spatial distribution of biomass, the distance between sources and sinks is crucial. Whereas short transport distances result in low transportation costs and, hence, in smaller conversion plants in a decentralised network structure, long transportation is used for the opposite situation. Due to economies of scale, larger plants might be favoured to smaller ones. The biomass value chain is characterised by the distribution of raw materials as biomass feedstock which are converted by multiple technologies into multiple output products such as bioenergy in electric and thermal form, biofuel, and biogas. The resulting decision levels of a biomass value chain from the biomass provision through the transport process to the conversion of biomass into the final product (electricity, heat, biogas, or biofuel) are illustrated in Figure 4.14:

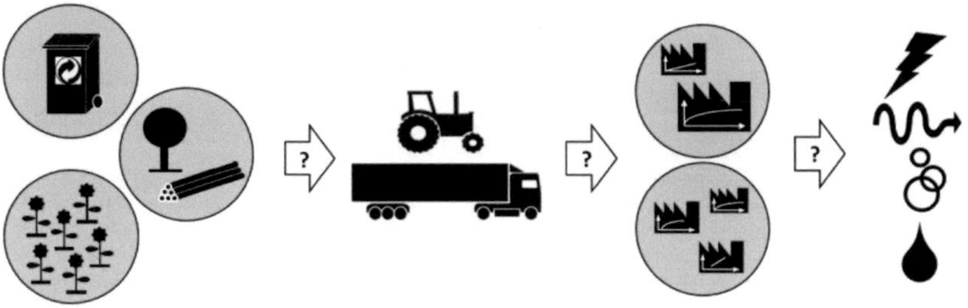

Figure 4.14: Decision levels of a biomass value chain.

4.2.1 Description of the methodology and model

According to the objective of RA2, the developed operations research model is based on a mathematical formulation, which maximises the profit as a function of revenue, depending on the output product and a certain market price, and the overall costs, which occur within the biomass value chain (i.e., biomass provision costs, transportation costs, and the investment related costs of the conversion technology).

The mathematical model is hereby characterised by a mixed-integer linear program. Various modelling approaches exist; however, for the requirement to incorporate multiple biomass feedstock and multiple conversion technologies, the application of linear models is obligatory. In the following the mathematical model formulation is presented starting with the sets, indices, decision variables, and parameters followed by the model equations. Subsequently, the implementation of the model is described briefly and important input data is stated. The specific parameters, i.e., the cost drivers of a biomass value and the input data for the model, are shown in Figure 4.15:

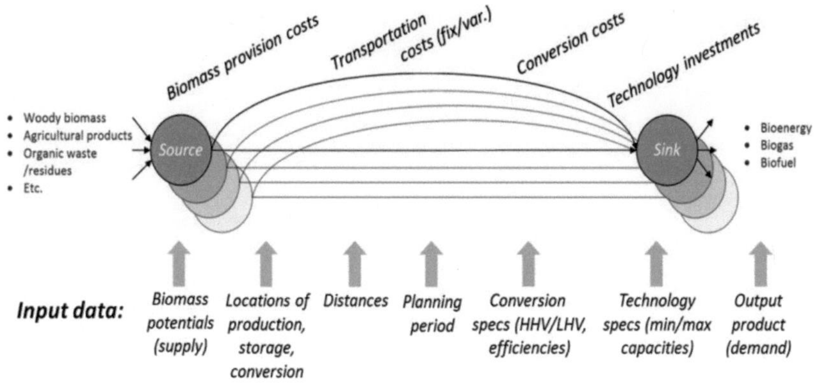

Figure 4.15: Parameter and model input data illustration.

4.2.1.1 Mathematical model formulation

Sets		Indices	
U	Set of biomass feedstock	$u \in U$	Biomass feedstock
T	Set of conversion technology	$t \in T$	Conversion technology
P	Set of output products	$p \in P$	Output product
N	Set of approximation interval	$n \in N$	Approximation interval (=6)
K	Set of plant per location	$k \in K$	Plant per location (≤10)
E	Set of edges	$(i,j) \in V$	Location nodes
V	Set of nodes	$(i,j) \in E$	Connection edges

Decision variables

$x_{uijk} \in \mathbb{R}^+$	Transported quantity of biomass feedstock *u* on arc (*i,j*) for facility *k*; $[t]$
$y^x_{uijk} \in \{1,0\}$	Binary variable representing the transported biomass according to x_{uijk}; $y^x_{uijk} = \{^{1;if\ biomass\ is\ transported\ on\ arc\ (i,j)}_{0;else}$
$e^1_{utpjk} \in \mathbb{R}^+$	Potentiated output of product *p* utilising biomass feedstock *u* with conversation technology *t* at sink *j* (plant locations) with facility *k*; $[MW]$, $[Nm^3]$, $[l]$
$e^2_{utpjk} \in \mathbb{R}^+$	Generated output of product *p* utilising biomass feedstock *u* with conversation technology *t* at sink *j* with facility *k*; $[MW]$, $[Nm^3]$, $[l]$
$e^{2d}_{utpjk} \in \mathbb{R}^+$	Dummy variable;
$y^e_{utpjk} \in \{1,0\}$	Binary variable representing the output supply according to e^1_{utpjk}; $y^e_{utpjk} = \{^{1;if\ output\ of\ product\ p\ is\ generated}_{0;else}$
$b_p \in \mathbb{R}^+$	Demand satisfaction variable of product *p*;
$n^{lt1e}_{utpjkn} \in \mathbb{R}^+$	Interval variable for the approximation of the capacity investments dependent on biomass feedstock *u* applying conversion technology *t* generating product *p* at sink *j* and facility *k* of interval *n*;
$n^{mt1e}_{utpjkn} \in \{1,0\}$	Binary variable for the capacity investment approximation in accordance to n^{lt1e}_{utpjkn}; $n^{mt1e}_{utpjkn} = \{^{1;if\ energy\ is\ generated}_{0;else}$
$y^a_{utpjk} \in \{1,0\}$	Binary variable for the technology assignment; $y^a_{utpjk} = \{^{1;if\ e^2_{utpjk} \neq 0}_{0;else}$
$y^b_{utpjk} \in \{1,0\}$	Binary dummy variable for the technology assignment; $y^b_{utpjk} = \{^{1;if\ e^{2d}_{utpjk} \neq 0}_{0;else}$

Model parameters

d_{ij}	Distance between source *i* and sink *j*; $[km]$
A_{ui}	Supply provided of biomass feedstock *u* at source location *i*; $[t]$
B_p	Output demand requested of product *p*; $[MW]$, $[Nm^3]$, $[l]$
$hours$	Planning period (= 3.600 hours); $[h]$
M	Big M – large number;
lf_j	Location factor of location *j*;
τ_j	Operating time of facilities at location *j* per planning period; $[h]$
η_{utp}	Efficiency of technology *t* converting biomass feedstock *u* into product *p*;
lhv_{utp}	Lower heating value; $\left[\frac{MJ}{t}\right]$
n_{tpn}^{1t}	Interval for capacity investment approximation;
$cap_u^{min^x}$	Minimal capacity for the transportation of biomass *u*; $[t]$
cap_{tp}^{min}	Minimal capacity for output generation by conversion technology *t*; $\left[\frac{MW}{t}\right]$, $\left[\frac{Nm^3}{t}\right]$, $\left[\frac{l}{t}\right]$
cap_{tp}^{max}	Maximal capacity for output generation by conversion technology *t*; $\left[\frac{MW}{t}\right]$, $\left[\frac{Nm^3}{t}\right]$, $\left[\frac{l}{t}\right]$
$cap_{tp}^{min^{pot}}$	Minimal potentiated capacity in accordance to cap_{tp}^{min}; $\left[\frac{MW}{t}\right]$, $\left[\frac{Nm^3}{t}\right]$, $\left[\frac{l}{t}\right]$
$cap_{tp}^{max^{pot}}$	Maximal potentiated capacity in accordance to cap_{tp}^{max}; $\left[\frac{MW}{t}\right]$, $\left[\frac{Nm^3}{t}\right]$, $\left[\frac{l}{t}\right]$
c_t^{om}	Operation and maintenance cost of conversion technology *t*; [*€ per output unit*]
c_t^{inv}	Investments of technology *t*; [*€ per output unit*]

c_{ut}^{invf}	Investment factor of technology *t* generating output *p*;
c_{ut}	Cost of biomass by-product disposal using feedstock *u* with technology *t*; $\left[\frac{€}{t}\right]$
c_{ut}^{f}	Cost ratio of biomass by-product disposal using feedstock *u* with technology *t*; [%]
c_u	Processing costs of biomass feedstock *u* per planning period; [€]
c_t	Cost of drying factor of technology *t*; $\left[\frac{€}{t}\right]$
c_u^{tf}	Fixed transportation costs of biomass feedstock *u*; [€]
c_u^{tv}	Variable transportation costs of biomass feedstock *u*; $\left[\frac{€}{km}\right]$
π_p	Price of output product *p*; $\left[\frac{€}{MW}\right]$; $\left[\frac{€}{Nm^3}\right]$; $\left[\frac{€}{l}\right]$
ε_p	Conversion factor of output product *p*;
$g2n_t^f$	Gross-to-net output generation factor of conversion technology *t*;

Model equations

Objective function:

$$Max \sum_{u \in U} \sum_{t \in T} \sum_{p \in P} \sum_{(i,j) \in E} \sum_{k \in K} \begin{pmatrix} \pi_p * e_{utpjk}^1 * \tau_j \\ -\left(c_u + c_t + c_{ut} * c_{ut}^f\right) * x_{uijk} \\ -\left(c_u^{tf} + c_u^{tv} * d_{ij}\right) * x_{uijk} \\ -\left(c_t^{inv} + c_t^{om}\right) * c_{ut}^{invf} * lf_j * e_{utpjk}^1 \end{pmatrix} \quad (1)$$

The objective function maximises the profit, which consists of the revenue of output product selling subtracted by the cost of biomass provision, transportation, and processing.

Supply provision:

$$\sum_{j\in E;k\in K} x_{uijk} \leq A_{ui} \qquad \forall\, u \in U \quad \forall\, i \in V \tag{2}$$

Biomass feedstock locations (sources) provide a certain amount of supply (A_{ui}).

Demand satisfaction per product:

$$\sum_{u\in U;t\in T;j\in J;k\in K} e^2_{utpjk} * g2n^f_t = b_p \qquad \forall\, p \in P \tag{3}$$

The demand of the final output product (b_p) is satisfied by the generated output ($\sum_{p\in P} e^2_{utpjk}$) and the technology specific output gross-to-net conversion ($g2n^f_t$).

Demand satisfaction restriction per product:

$$b_p * \varepsilon_p \geq B_p \qquad \forall\, p \in P \tag{4}$$

Product specific demand is satisfied (B_p) by converting the generated output ($b_p * \varepsilon_p$).

Demand minimum and maximum restrictions:

$$b_p \geq B^{min};\ b_p \leq B^{max} \qquad \forall\, p \in P \qquad (5)\ (6)$$

Binary biomass flow (bundle constraint) and minimal biomass capacity:

$$x_{uijk} \leq y^x_{uijk} * M \qquad \forall\, u \in U \tag{7}$$

$$\forall\, k \in K$$

$$x_{uijk} \geq y^x_{uijk} * cap^{min^x}_u \qquad \forall\, i,j \in V \tag{8}$$

If biomass is transported (x_{uijk}), a minimal amount of biomass is required ($cap^{min^x}_u$).

Binary output flow (bundle constraint) and minimal output capacity:

$$e^1_{utpjk} \leq y^e_{utpjk} * M \qquad (9)$$

$$e^1_{utpjk} \geq y^e_{utpjk} * cap^{min^{pot}}_{tp} \qquad (10)$$

$$\forall u \in U \quad \forall t \in T \quad \forall p \in P \quad \forall j \in V \quad \forall k \in K$$

A minimal output capacity ($cap^{min^{pot}}_{tp}$) for the output generating (e^1_{utpjk}) is required.

Equations for the approximation of the capacity investment function

Transport and output equilibrium weighting approximation:

$$\sum_{i \in E} \frac{x_{uijk} * \eta_{utp} * lhv_{utp}}{\tau_j * hours} = e^2_{utpjk} + e^{2d}_{utpjk} \qquad \forall u \in U \quad \forall t \in T \quad \forall p \in P \quad \forall j \in V \quad \forall k \in K \qquad (11)$$

Depending on the transported biomass (x_{uijk}) and the applied conversion technology, the equilibrium constraint approximates the capacity degression with two modelling variables whereas variable e^2_{utpjk} represents the approximated output and e^{2d}_{utpjk} a dummy variable.

Biomass and output alignment constraints:

$$e^2_{utpjk} \leq y^a_{utpjk} * M \qquad (12)$$

$$e^{2d}_{utpjk} \leq y^b_{utpjk} * M \qquad (13)$$

$$y^a_{utpjk} + y^b_{utpjk} = 1 \qquad (14)$$

$$\forall u \in U \quad \forall t \in T \quad \forall p \in P \quad \forall j \in V \quad \forall k \in K$$

To align the biomass type and the technology as well as the output product, constraint (14) restricts the binary variables y^a_{utpjk} and y^b_{utpjk} while affecting constraint (11) to consider only relevant biomass and technology combinations. The bundle constraints (12) and (13) apply the Big M method to formulate constraint (14).

Potentiated output product approximation:

The approach of linearising the investment function and defining the capacity intervals of the conversion technology according to the Multiple-Choice methodology applied by Frank Schwaderer (2012), who approximated the step-wise linear function based on a SOS2-implementation (Special-Order-Sets of type 2). Herein, the degression function applies a specific technology exponent (see Table 4.10).

$$e^{1}_{utpjk} = \sum_{n \in N} n^{1t}_{tpn} * n^{lt1e}_{utpjkn} * cap_{tp}^{max^{pot}} \qquad \begin{matrix} \forall u \in U \\ \forall t \in T \\ \forall p \in P \\ \forall j \in V \\ \forall k \in K \end{matrix} \qquad (15)$$

Based on the potentiated capacity the output (e^{1}_{utpjk}) is defined.

Output product approximation:

$$e^{2}_{utpjk} = \sum_{n \in N} n^{1t}_{tpn} * n^{lt1e}_{utpjkn} * cap_{tp}^{max} \qquad \begin{matrix} \forall u \in U \\ \forall t \in T \\ \forall p \in P \\ \forall j \in V \\ \forall k \in K \end{matrix} \qquad (16)$$

Based on the capacity degression the output (e^{2}_{utpjk}) is defined

Approximation limit with biomass feedstock consideration:

$$n^{lt1e}_{utpjkn} \leq n^{mt1e}_{utpjkn} \qquad \begin{matrix} \forall u \in U \\ \forall t \in T \\ \forall p \in P \\ \forall j \in V \\ \forall k \in K \\ \forall n \in N; n \leq 2 \end{matrix} \qquad (17)$$

Approximation neighbour with biomass feedstock consideration:

$$n^{lt1e}_{utpjkn} \leq n^{mt1e}_{utpjkn-1} + n^{mt1e}_{utpjkn} \qquad \begin{matrix} \forall u \in U \\ \forall t \in T \\ \forall p \in P \\ \forall j \in V \\ \forall k \in K \\ \forall n \in N; n > 2 \end{matrix} \qquad (18)$$

$$n_{utpjkn}^{lt1e} \leq n_{utpjkn-1}^{mt1e} \qquad \begin{array}{l} \forall u \in U \\ \forall t \in T \\ \forall p \in P \\ \forall j \in V \\ \forall k \in K \\ \forall n \in N \end{array} \tag{19}$$

Approximation interval with biomass feedstock consideration:

$$n_{utpjkn}^{lt1e} \leq 1 \qquad \begin{array}{l} \forall u \in U \\ \forall t \in T \\ \forall p \in P \\ \forall j \in V \\ \forall k \in K \\ \forall n \in N \end{array} \tag{20}$$

Approximation interval sum with biomass feedstock consideration:

$$\sum_{n} n_{utpjkn}^{lt1e} = 1 \tag{21}$$

$$\sum_{n} n_{utpjkn}^{mt1e} = 1 \tag{22}$$

$$\forall u \in U, \forall t \in T, \forall p \in P, \forall j \in V, \forall k \in K$$

4.2.1.2 Model implementation

The model is implemented in GAMS (General Algebraic Modelling System) and embedded within a decision support system. Following the principle of separation of logic and data, the decision support system is divided into the model and a database to allow independent data and model handling on the one hand, and to increase the application flexibility and functionality to guarantee dynamic modelling on the other hand (Rosenthal 2012, p.7). The database as well as the user interface are implemented in a MS Excel-based environment to ease the use. The system overview covering the data flows between MS Excel and GAMS is shown in Figure 4.16. After input data maintenance, the user runs the model. The subsequent processes are automatically executed in the background and are explained in the following.

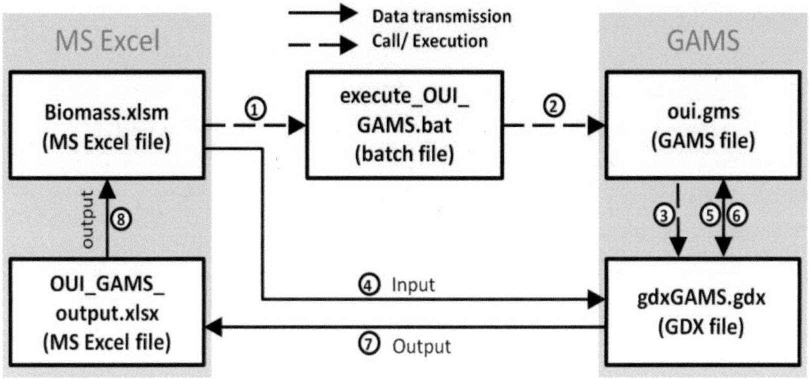

Figure 4.16: System overview.

1. The user executes the batch file with a macro from the user interface in MS Excel.
2. The batch file calls, depending on the location of the GAMS directory, the GAMS model for execution.
3. GAMS calls the GDX file to obtain required model data from the database in MS Excel.
4. GDX reads the input data.
5. GDX forwards input data to the GAMS model.
6. After model execution, GAMS returns model output results to GDX.
7. GDX writes model output results into MS Excel output file.
8. The main MS Excel file reads and presents the model output results.

While the data is processed by GAMS, the user interface as well as MS Excel is blocked for further entry.

4.2.1.3 Input data definition

The crucial input data are the various biomass potentials, which define the type of biomass feedstock applied. Depending on the available biomass feedstock the type of conversion technology is assigned. The observed biomass feedstock is: forest residues; household waste and green waste; straw; vineyard residues such as prunings, pomace, and yeast; manure; crop residues and woody biomass. The integrated conversion technologies with the specific output product are: anaerobic digestion (AD) for biogas, waste-to-energy (WtE) for electric energy, combustion (Comb) for electric (el.) and thermal (ther.) energy, gasification with a downdraft gasifier (DG), a fluidised bed gasifier (FBG), and the biomass integrated gasification combined cycle (BIGCC) producing electric energy; and biofuel conversion of cereals (BeCereals) and woody (BeLigno) materials into bioethanol. The according mapping of biomass feedstock and conversion technology as well as the corresponding output product is shown in the following matrix (Table 4.7):

Table 4.7: Biomass feedstock to conversion technology mapping matrix.

	AD [Nm^3]	**WtE** [MW]	**Comb** [MW]	**DG** [MW]	**FBG** [MW]	**BIGCC** [MW]	**Be Cereals** [l]	**Be Ligno** [l]
Cereals							biofuel	
Household waste		electric						
Green waste	biogas							
Manure	biogas							
Forest residues			el./ther.	electric	electric	electric		biofuel
Straw			el./ther.	electric	electric	electric		biofuel
Vre prun-ings			el./ther.	electric	electric	electric		biofuel

Further crucial input data is the assumed lower heating value (LHV) of the different biomass feedstock. The LHV depends on the biomass and the applied conversion technology, which is linked to the output product, and is the main parameter next to the efficiency (η) that affects the conversion process. With reference to the reports provided by RA3 (Bidart et al. 2014-2015) and additional literature review, the following data has been used (Table 4.8):

Table 4.8: Lower Heating Values (LHV) and technology efficiency (η) (Source: RA3-Reports 1, 3, and 4).

	LHV [MJ/t]	η [%] (technology and output dependent)
Cereals	15 082	100
Household waste	15 000	21
Green waste	10.541	42.61
Manure	12.42	42.61
Forest residues	7 959	25.17 | 29.18 | 29.34 | 40.6 | 100
Straw	14 200	25.17 | 29.18 | 29.34 | 40.6 | 100
Vre prunings	14 200	25.17 | 29.18 | 29.34 | 40.6 | 100

For the by-product disposal (e.g., removing the ash residues of combustion and gasification processes) (c_{ut}, c_{ut}^{f}), the cost of biomass feedstock (c_u), and the fixed (c_u^{tf}) and variable (c_u^{tv}) cost parameters are (Table 4.9):

Table 4.9: Biomass-specific input parameters (Source: own calculations and www.agrarheute.com).

	c_u $[\frac{€}{t}]$	c_{ut} $[\frac{€}{t}]$	c_{ut}^{f} [%]	c_u^{tf} $[\frac{€}{t}]$	c_u^{tv} $[\frac{€}{t*km}]$
Cereals	150	-	-	1.55	0.18
Household waste	30	70	8	2.5	0.09
Green waste	50	-	-	1.55	0.18
Manure	30	-	-	1.55	0.18
Forest residues	50	30	9	1.55	0.18
Straw	70	30	30	1.55	0.52
Vre prunings	60	30	30	1.55	0.18

In accordance to the biomass feedstock and conversion technology mapping matrix, the technology-specific parameters are as follows (Table 4.10):

Table 4.10: Techonology parameters (Source: RA3-Reports)

	AD [Nm3]	**WtE** [MW]	**Comb** [MW]	**DG** [MW]	**FBG** [MW]	**BIGCC** [MW]	**Be Cereals** [l]	**Be Ligno** [l]
c_t^{om}	0.12	0.04	0.047	0.045	0.045	0.045	0.07	0.07
c_t	-	-	-	0.08	0.08	0.08	-	-
$g2n_t^f$	0.9	-	-	-	-	-	-	-
$c_{ut}^{inv^f}$	17 449	4 258 717	3 679 014	2 742 590	5 887 806	11 324 452	1.037	1.154
cap_{tp}^{min}	10	38.5	3	0.05	1.8	6	55	10
cap_{tp}^{max}	1100	58.625	25	2	15	250	150	250
Techn. exponent	0.859	0.820	0.547	0.808	0.286	0.577	1	0
$cap_{tp}^{min^{pot}}$	7.22	19.956	1.824	0.089	1.183	2.811	55	2
$cap_{tp}^{max^{pot}}$	408.64	28.173	5.817	1.75	2.172	24.162	150	6

Moreover, the technology investments are based on an average cost parameter ($\boldsymbol{c_t^{inv}}$) defining additional portion of expenses for all technologies to be 0.155 (see Schwaderer 2012). The location factor $\boldsymbol{lf_j}$ is assumed to be 0.98 for France as well as 0.92 for Germany and Switzerland, and was provided by RA3. The distances $\boldsymbol{d_{ij}}$ between the locations are based on own calculations spanning a 36x36 matrix with a distance for $\boldsymbol{i} = \boldsymbol{j}$ to be 5 km (source and sink location are the same). The locations $(\boldsymbol{i}, \boldsymbol{j})$ were defined as biomass sources and biomass conversion facilities (see Table 4.11). The assumptions for the preselection of locations are the proximity to infrastructure such as roads and rivers as well as cities.

Table 4.11: Biomass production and potential conversion locations in the URR.

ID	District / Kanton / Arrondissement	City
1	Karlsruhe District	Bruchsal
2	Karlsruhe City	Ettlingen
3	Rastatt	Gaggenau
4	Baden-Baden	Geroldsau
5	Ortenaukreis	Appenweier
6	Ortenaukreis	Haslach
7	Emmendingen	Denzlingen
8	Freiburg	BadKrozingen
9	Breisgau	Titisee
10	Lörrach	Zell
11	Waldshut	Haeusern
12	Aargau	Frick
13	Aargau	Wohlen
14	Solothurn	Balsthal
15	Basel	Sissach
16	Jura	Delemont
17	Jura	Porrentruy
18	Altkirch	Altkirch
19	Mulhouse	Wittenheim
20	Thann	Thann
21	Guebwiller	Meyenheim
22	Colmar	Wolfgantzen
23	Colmar	Munster
24	Ribeauville	SMM
25	Selestat	Marckolsheim
26	Selestat	Barr
27	Molsheim	Urmatt
28	Straßburg	Fegersheim
29	Straßburg	Brumath
30	Saverne	Saverne
31	Haguenau	Soufflenheim
32	Wissemburg	Reichshoffen
33	Germersheim	Hoffen
34	Landau	Kandel
35	Südl. Weinstraße	BadBergzabern
36	Südwestpfalz	Dahn

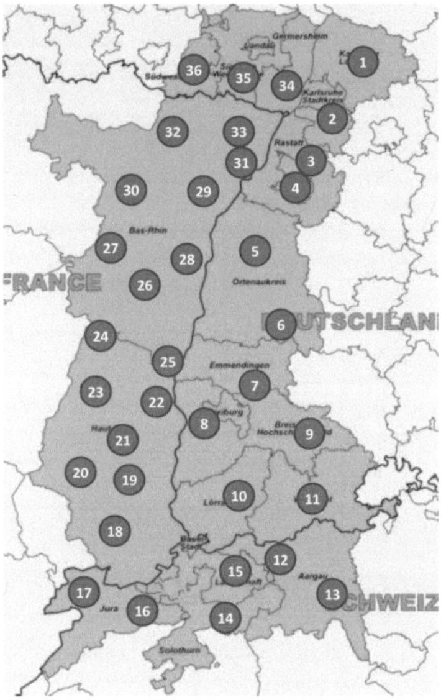

Figure 4.17: Biomass production and potential conversion locations in the URR.

The interval steps for the piecewise linear function consist of 1 to 6 steps and the interval capacity investment approximation (n_{tpn}^{1t}) is calculated according to the technology exponent and the investment factor (c_{tp}^{invf}). By taking the following scenario analysis into account the price (π_p) as well as the demand parameter (B_p) are set in accordance to the requirement to utilise the complete biomass feedstock potentials. Hence, the biomass price and output product demand are increased to enable the entire conversion of available biomass feedstock.

4.2.2 Results of the model for currently unused biomass potentials

Various unused biomass potentials exist in the Upper Rhine Region (URR) as stated in section 4.1. However, not all biomass potentials can be utilised due to certain technology capacity restrictions and economic reasons. For instance, the current potential amount of green waste to be used for anaerobic digestion in the URR is not sufficient to be converted into biogas for thermal or upgrading purposes. The same accounts for vineyard residues such as pomace and yeast. Even a combined utilisation of all three types of biomass is not sufficient for biogas production. Therefore, only the introduced types of biomass feedstock above (i.e., cereals, household waste, manure, forest residues, straw, vineyard residues prunings) are applied in the following scenario analysis. The scenario analysis starts with possible biomass utilisation pathways according to the biomass potentials in 2010 and the 2030 BAU, ConRec and MaxEx scenarios (see section 4.1). Subsequently, biogas potentials in 2030 for the three scenarios are analysed and evaluated. Additionally, the number of facilities at every possible location is restricted to one ($k = 1$). The reason is the increased computational time in case of more than one facility. However, for certain scenarios this restriction is disabled to investigate the effects on the network structure. For instance, if the technological capacity is satisfied at one facility location, another facility at another location is created leading to a decentralised network. To enable the creation of more than one facility at one location, the parameter k is increased and its effects are investigated.

Scenario results of forest residues

The available forest residues potential for the use of electric and thermal energy by applying combustion and gasification technologies or the production of bioethanol is approximately 163 kt_{atro}/a based on the reference year 2010. By applying the model to obtain the most economic utilisation of forest residues the results of transported forest residues for bioenergy generation are:

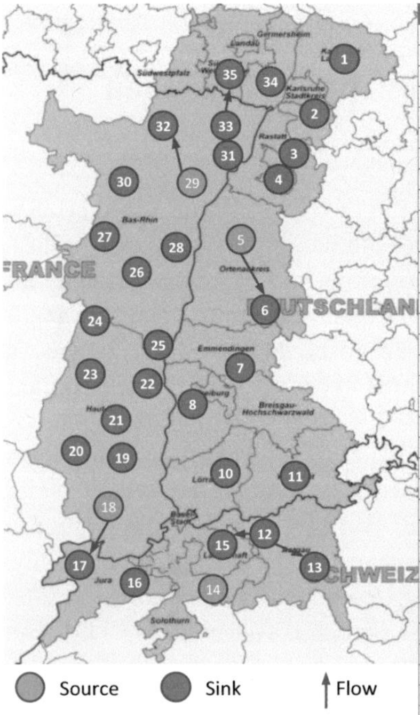

ID	Location
01	Bruchsal
02	Ettlingen
03	Gaggenau
04	Geroldsau
05	Appenweier
06	Haslach
07	Denzlingen
08	BadKrozingen
09	Titisee
10	Zell
11	Haeusern
12	Frick
13	Wohlen
14	Balsthal
15	Sissach
16	Delemont
17	Porrentruy
18	Altkirch
19	Wittenheim
20	Thann
21	Meyenheim
22	Wolfgantzen
23	Munster
24	SMM
25	Marckolsheim
26	Barr
27	Urmatt
28	Fegersheim
29	Brumath
30	Saverne
31	Soufflenheim
32	Reichshoffen
33	Hoffen
34	Kandel
35	BadBergzabern
36	Dahn

Output data of scenario:

Forest residues 2010

Biomass utilised:
163 kt_{atro}/a

Number of facilities:
31

Total distance:
336 km

Total transport costs:
0.5 Mio.€

Total biomass provision costs:
43 Mio.€

Total investments:
9.5 Mio.€

Total energy:
18.5 MW_{el} (DG)

Figure 4.18: Scenario results of forest residues 2010.

The forest residues biomass potentials can be utilised for energy production of around 18.5 MW_{el} via DG gasification with a conversion capacity range of 0.12-1.3 MW_{el}. The average energy output is 0.6 MW_{el} with average investments of about 0.3 Mio.€ per facility.

The combustion and FBG/BIGCC gasification as well as the conversion into bioethanol is due to the restricted amount of biomass limited. Although technologies with greater capacity ranges are available for conversion, only small scale gasification facilities are applied forming a decentralised network structure.

A comparison of gasification technologies is provided by Kerdoncuff (2008, p. 37), who highlights that a great amount of biomass is required for high capacity facilities. By forcing the model into a centralised structure and removing the DG option, the facility locations 1 (Bruchsal) and 18 (Altkirch) are chosen for FBG gasification with a total distance of 2 664 km, transport cost of 2.5 Mio.€ and investments of 3.4 Mio.€ generating only a fraction of the initial bioenergy.

Scenario results of household waste

The available household waste potentials for the use of electric and thermal energy by applying waste-to-energy technologies is approximately 875 t/a based on the reference year 2010. By applying the model, the results of transported household waste for bioenergy generation are:

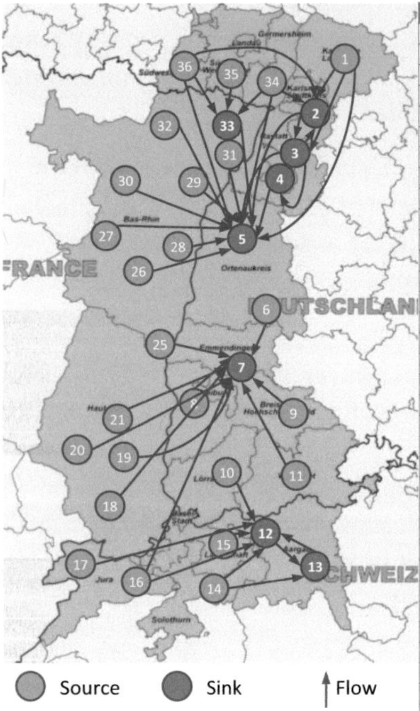

ID	Location
01	Bruchsal
02	Ettlingen
03	Gaggenau
04	Geroldsau
05	Appenweier
06	Haslach
07	Denzlingen
08	BadKrozingen
09	Titisee
10	Zell
11	Haeusern
12	Frick
13	Wohlen
14	Balsthal
15	Sissach
16	Delemont
17	Porrentruy
18	Altkirch
19	Wittenheim
20	Thann
21	Meyenheim
22	Wolfgantzen
23	Munster
24	SMM
25	Marckolsheim
26	Barr
27	Urmatt
28	Fegersheim
29	Brumath
30	Saverne
31	Soufflenheim
32	Reichshoffen
33	Hoffen
34	Kandel
35	BadBergzabern
36	Dahn

Output data of scenario:

Household waste 2010

Biomass utilised:
841 t/a

Number of facilities:
3 (7)

Total distance:
2 343 km

Total transport costs:
5.9 Mio.€

Total biomass provision costs:
30 Mio.€

Total investments:
54 Mio.€

Total energy:
70.7 MW_{el} (WtE)

Figure 4.19: Scenario results of household waste 2010.

The household waste biomass potentials can be utilised for energy production of around 70.7 MW_{el} with a conversion capacity range of 20-28 MW_{el}. The average energy output is about 23 MW_{el} with average investments of 23.6 Mio.€ per facility. Three big facilities, namely 5 (Appenweier) with 28 MW_e, 7 (Denzlingen) with 20 MW_{el} and 12 (Frick) with 23 MW_{el}, generate 99% of the total bioenergy and are located on the eastern side of the URR forming a centralised structure.

Scenario results of straw

Straw biomass potentials for bioenergy usage in 2010 were 930 kt/a. This amount can be combusted, gasified, or converted into bioethanol. The model results for straw conversion are:

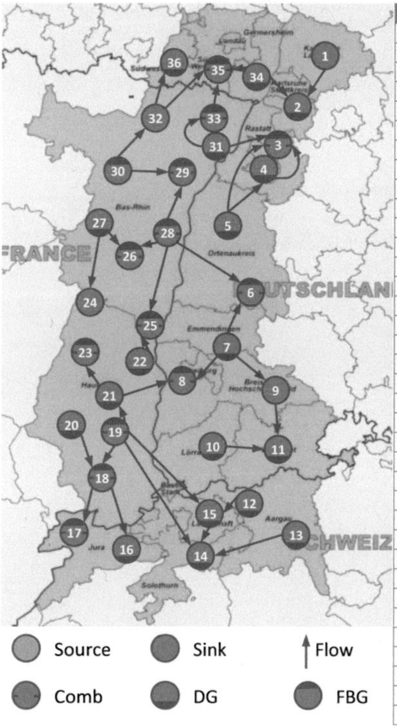

ID	Location
01	Bruchsal
02	Ettlingen
03	Gaggenau
04	Geroldsau
05	Appenweier
06	Haslach
07	Denzlingen
08	BadKrozingen
09	Titisee
10	Zell
11	Haeusern
12	Frick
13	Wohlen
14	Balsthal
15	Sissach
16	Delemont
17	Porrentruy
18	Altkirch
19	Wittenheim
20	Thann
21	Meyenheim
22	Wolfgantzen
23	Munster
24	SMM
25	Marckolsheim
26	Barr
27	Urmatt
28	Fegersheim
29	Brumath
30	Saverne
31	Soufflenheim
32	Reichshoffen
33	Hoffen
34	Kandel
35	BadBergzabern
36	Dahn

Output data of scenario:

Straw 2010

Biomass utilised: 930 kt/a

Number of facilities: 66

Total distance: 1 522 km

Total transport costs: 9.5 Mio.€

Total biomass provision costs: 330 Mio.€

Total investments: 133 Mio.€

Total energy: 96.4 MW_{el} (Comb/DG/ FBG) 85.7 MW_{th} (Comb)

Figure 4.20: Scenario results of straw 2010.

The straw biomass potentials can be utilised for energy production of around 96.4 MW_{el} and 85.7 MW_{th} with a conversion capacity range of 1.3-1.7 MW_{el} for gasification and 1.8-5.5 MW_{th} for combustion. The average energy output is 1.7 MW_{el} for gasification and 4 MW_{th} for combustion with average investments of around 2 Mio.€ per facility. In total, 30 combustion and 36 gasification facilities (21 DG, 15 FBG) are distributed forming a decentralised network structure with mostly local utilisation and only little biomass distribution. Mainly because of the limited capacity and the restriction to allow only one conversion facility per location, transportation of biomass occurs.

Scenario results of vineyard residues (prunings)

The total vineyard residues (prunings) potential in 2010 is approximately 151 t/a and can be converted like woody biomass into thermal or electric bioenergy. The model results for the woody vineyard residues conversion are:

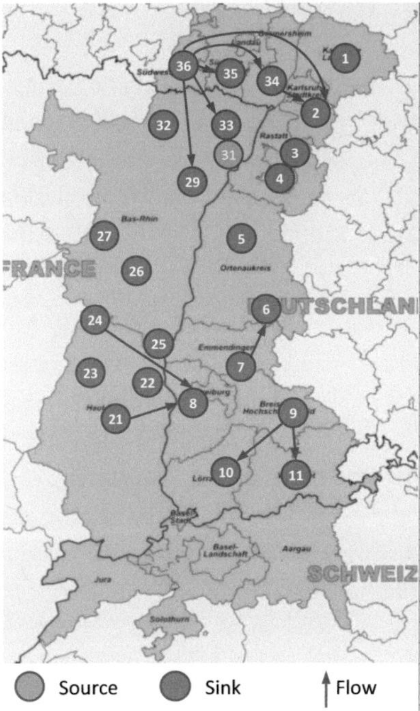

ID	Location
01	Bruchsal
02	Ettlingen
03	Gaggenau
04	Geroldsau
05	Appenweier
06	Haslach
07	Denzlingen
08	BadKrozingen
09	Titisee
10	Zell
11	Haeusern
12	Frick
13	Wohlen
14	Balsthal
15	Sissach
16	Delemont
17	Porrentruy
18	Altkirch
19	Wittenheim
20	Thann
21	Meyenheim
22	Wolfgantzen
23	Munster
24	SMM
25	Marckolsheim
26	Barr
27	Urmatt
28	Fegersheim
29	Brumath
30	Saverne
31	Soufflenheim
32	Reichshoffen
33	Hoffen
34	Kandel
35	BadBergzabern
36	Dahn

Output data of scenario:

Vineyard residues (prunings) 2010

Biomass utilised:
151 kt/a

Number of facilities:
24

Total distance:
576 km

Total transport costs:
0.7 Mio.€

Total biomass provision costs:
46 Mio.€

Total investments:
16 Mio.€

Total energy:
30.57 MW_{el} (DG)

Figure 4.21: Scenario results of vineyard residues (prunings) 2010.

The entire utilisation of woody vineyard residues (prunings) results in a generation of 30.57 MW_{el} of bioenergy with a conversion capacity range of 0.9-1.75 MW_{el} for low-scale gasification. The average energy output is 1.27 MW_{el} with average investments of about 0.67 Mio.€ per facility. The biomass distribution structure is divers due to the heterogeneous biomass potentials emphasising one region in the North and one in the mid-area of the URR. The fact that at location 36 (Dahn) the biomass potential exceeds the conversion capacity and biomass needs to be transported to other facility locations (i.e., 29, 33, 34, 35) highlights the Palatinate wine region.

4.2.3 Results of the model for future biomass use

Whereas the previous subsection focuses on the unused biomass potentials, the following section investigates potential biomass utilisation pathways in the year 2030 in accordance to the three scenarios: Business-As-Usual (BAU), Conservation and Recreation (ConsRec), Maximal Exploitation (MaxEx). Those scenarios are described in section 4.1.3 of this scientific report.

At first, the model results of the BAU scenarios are described for the following biomass feedstock: forest residues, straw, and vineyard residues (prunings). Due to the specific scenario assumptions the amount of biomass potentials is reduced limiting its utilisation by the available conversion technologies. Hence, the provided quantity of manure, household, and green waste, as well as crop residues and woody biomass conversion for biogas production is not sufficient in the BAU scenarios. Only the amount of manure is relevant for biogas production in the MaxEx scenario, but solely one location (Frick; 12) provides enough biomass for the conversion of 92 kt/a into 24 Nm^3 of biogas at investments of 0.1 Mio.€. Woody biomass in addition only satisfies the minimal capacities in the MaxEx scenario.

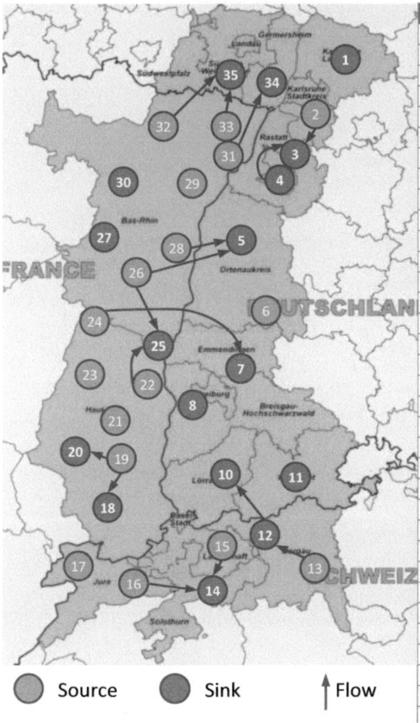

ID	Location
01	Bruchsal
02	Ettlingen
03	Gaggenau
04	Geroldsau
05	Appenweier
06	Haslach
07	Denzlingen
08	BadKrozingen
09	Titisee
10	Zell
11	Haeusern
12	Frick
13	Wohlen
14	Balsthal
15	Sissach
16	Delemont
17	Porrentruy
18	Altkirch
19	Wittenheim
20	Thann
21	Meyenheim
22	Wolfgantzen
23	Munster
24	SMM
25	Marckolsheim
26	Barr
27	Urmatt
28	Fegersheim
29	Brumath
30	Saverne
31	Soufflenheim
32	Reichshoffen
33	Hoffen
34	Kandel
35	BadBergzabern
36	Dahn

Output data of scenario:

Forest residues BAU

Biomass utilised:
14.869 kt_{atro}/a

Number of facilities:
16

Total distance:
639 km

Total transport costs:
0.06 Mio.€

Total biomass provision costs:
3.88 Mio.€

Total investments:
0.863 Mio.€

Total energy:
1.68 MW_{el} (DG)

Figure 4.22: Scenario results of forest residues BAU.

4.2.3.1 Results of the model for the BAU scenario

The forest residues biomass potentials of the BAU scenario is assumed to be approximately 14.869 kt_{atro} per year and can be utilised for energy production of around 1.68 MW_{el} via small-scale down draft (DG) gasification with a conversion capacity range of 0.08-0.13 MW_{el}. The average energy output is 0.1 MW_{el} with average investments of 0.055 Mio.€ per facility.

Due to the small biomass potentials distributed within the URR, the structure is scattered with only 16 small gasification facilities, which mostly apply the technology on-site with very little biomass transportation.

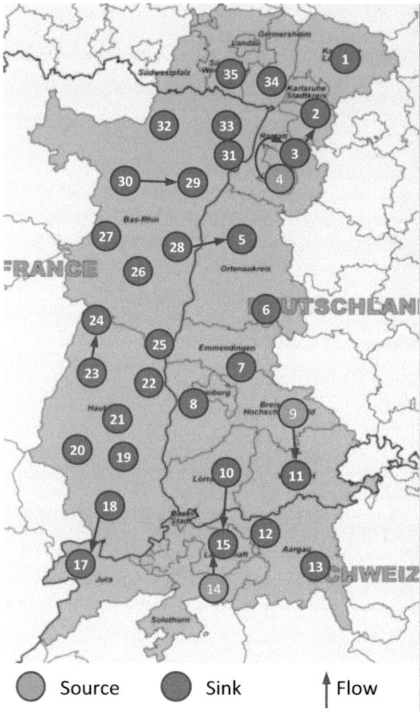

ID	Location
01	Bruchsal
02	Ettlingen
03	Gaggenau
04	Geroldsau
05	Appenweier
06	Haslach
07	Denzlingen
08	BadKrozingen
09	Titisee
10	Zell
11	Haeusern
12	Frick
13	Wohlen
14	Balsthal
15	Sissach
16	Delemont
17	Porrentruy
18	Altkirch
19	Wittenheim
20	Thann
21	Meyenheim
22	Wolfgantzen
23	Munster
24	SMM
25	Marckolsheim
26	Barr
27	Urmatt
28	Fegersheim
29	Brumath
30	Saverne
31	Soufflenheim
32	Reichshoffen
33	Hoffen
34	Kandel
35	BadBergzabern
36	Dahn

Output data of scenario:

Straw BAU

Biomass utilised:
93 kt/a

Number of facilities:
31

Total distance:
431 km

Total transport costs:
0.45 Mio.€

Total biomass provision costs:
33 Mio.€

Total investments:
10 Mio.€

Total energy:
18.7 MW_{el} (DG)

Figure 4.23: Scenario results of straw BAU.

The straw biomass potentials of the BAU scenario is assumed to be approximately 93 kt per year and can be utilised for energy production of around 18.7 MW_{el} via small-scale down draft (DG) gasification with a conversion capacity range of 0.09-0.18 MW_{el}. The average energy output is 0.6 MW_{el} with average investments of 0.32 Mio.€ per facility.

The resulting structure is characterised by an almost complete on-site utilisation of straw potentials applying small-scale DG gasification.

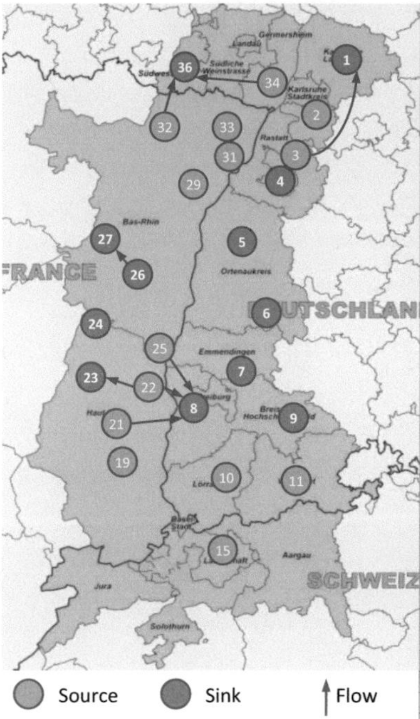

ID	Location
01	Bruchsal
02	Ettlingen
03	Gaggenau
04	Geroldsau
05	Appenweier
06	Haslach
07	Denzlingen
08	BadKrozingen
09	Titisee
10	Zell
11	Haeusern
12	Frick
13	Wohlen
14	Balsthal
15	Sissach
16	Delemont
17	Porrentruy
18	Altkirch
19	Wittenheim
20	Thann
21	Meyenheim
22	Wolfgantzen
23	Munster
24	SMM
25	Marckolsheim
26	Barr
27	Urmatt
28	Fegersheim
29	Brumath
30	Saverne
31	Soufflenheim
32	Reichshoffen
33	Hoffen
34	Kandel
35	BadBergzabern
36	Dahn

Output data of scenario:

Vineyard residues (prunings) BAU

Biomass utilised: 14.3 kt/a

Number of facilities: 11

Total distance: 370 km

Total transport costs: 0.053 Mio.€

Total biomass provision costs: 4.3 Mio.€

Total investments: 1.5 Mio.€

Total energy: 2.86 MW_{el} (DG)

Figure 4.24: Scenario results of vineyard residues (prunings) BAU.

The prunings potentials of vineyard residues of the BAU scenario are assumed to be approximately 14.3 kt per year and can be utilised for energy production of around 2.86 MW_{el} via small-scale down draft (DG) gasification with a conversion capacity range of 0.09-0.88 MW_{el}. The average energy output is 0.26 MW_{el} with average investments of about 0.13 Mio.€ per facility.

Only small amounts of biomass feedstock are distributed. On-site gasification of vineyard residues seems to be favourable. At certain facility locations the potentials are too small for conversions (e.g., 10, 11, 15, etc.) and the collection of biomass via transportation is not required.

Results of the model for the ConsRec scenario

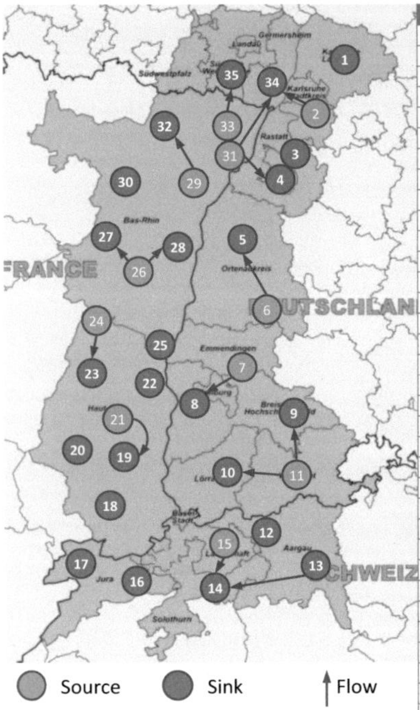

ID	Location
01	Bruchsal
02	Ettlingen
03	Gaggenau
04	Geroldsau
05	Appenweier
06	Haslach
07	Denzlingen
08	BadKrozingen
09	Titisee
10	Zell
11	Haeusern
12	Frick
13	Wohlen
14	Balsthal
15	Sissach
16	Delemont
17	Porrentruy
18	Altkirch
19	Wittenheim
20	Thann
21	Meyenheim
22	Wolfgantzen
23	Munster
24	SMM
25	Marckolsheim
26	Barr
27	Urmatt
28	Fegersheim
29	Brumath
30	Saverne
31	Soufflenheim
32	Reichshoffen
33	Hoffen
34	Kandel
35	BadBergzabern
36	Dahn

Output data of scenario:

Forest residues ConRec

Biomass utilised: 81.775 kt_{atro}/a

Number of facilities: 24

Total distance: 619 km

Total transport costs: 0.3 Mio.€

Total biomass provision costs: 21 Mio.€

Total investments: 4.7 Mio.€

Total energy: 9.23 MW_{el} (DG)

Figure 4.25: Scenario results of forest residues ConRec.

The forest residues biomass potential of the ConRec scenario is assumed to be approximately 81.775 kt_{atro} per year and can be utilised for energy production of around 9.23 MW_{el} via down draft (DG) gasification with a conversion capacity range of 0.08-0.8 MW_{el}. The average energy output is 0.38 MW_{el} with average investments of 0.2 Mio.€ per facility.

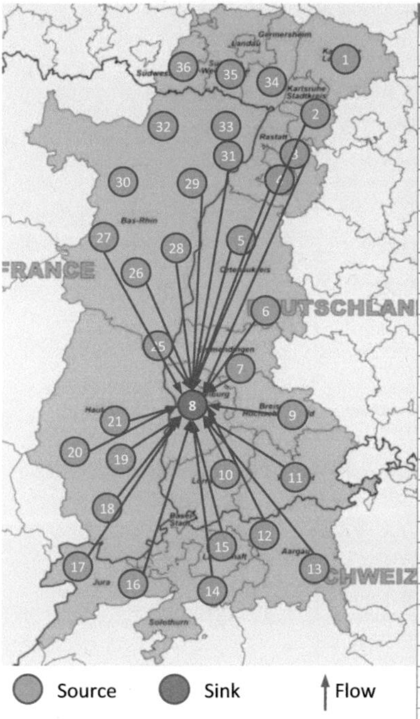

ID	Location
01	Bruchsal
02	Ettlingen
03	Gaggenau
04	Geroldsau
05	Appenweier
06	Haslach
07	Denzlingen
08	BadKrozingen
09	Titisee
10	Zell
11	Haeusern
12	Frick
13	Wohlen
14	Balsthal
15	Sissach
16	Delemont
17	Porrentruy
18	Altkirch
19	Wittenheim
20	Thann
21	Meyenheim
22	Wolfgantzen
23	Munster
24	SMM
25	Marckolsheim
26	Barr
27	Urmatt
28	Fegersheim
29	Brumath
30	Saverne
31	Soufflenheim
32	Reichshoffen
33	Hoffen
34	Kandel
35	BadBergzabern
36	Dahn

Output data of scenario:

Household waste ConRec

Biomass utilised: 335 kt/a

Number of facilities: 1

Total distance: 2 139 km

Total transport costs: 3.26 Mio.€

Total biomass provision costs: 12 Mio.€

Total investments: 21.5 Mio.€

Total energy: 28.2 MW_{el} (WtE)

Figure 4.26: Scenario results of household waste ConRec.

The waste-to-energy technology requires a high amount of substrates to reach the minimal capacity level. In the BAU scenario the amount of household waste was insufficient; in the ConsRec the total amount enables the establishment of one waste processing facility in the centre of the URR at Bad Kronzingen. The facilities capacity generates approximately 28 MW_{el} of electric energy with 21.5 Mio.€ of investments. Interesting is the fact that not the complete biomass potential of household waste is applied for conversion such as at facility locations 34, 35, 36 for instance; this is explained by the total utilisation of the waste-to-energy technology capacity.

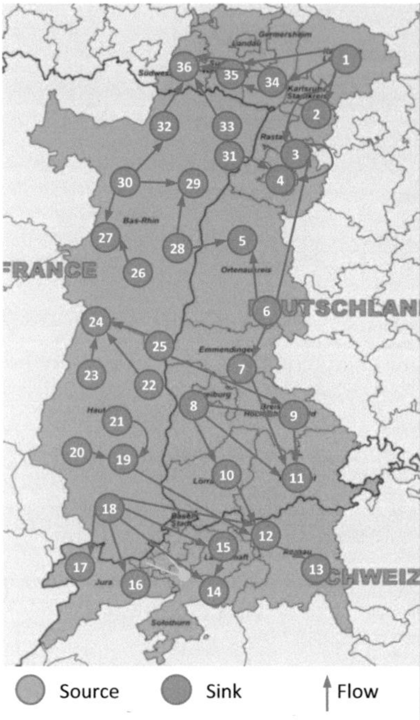

ID	Location
01	Bruchsal
02	Ettlingen
03	Gaggenau
04	Geroldsau
05	Appenweier
06	Haslach
07	Denzlingen
08	BadKrozingen
09	Titisee
10	Zell
11	Haeusern
12	Frick
13	Wohlen
14	Balsthal
15	Sissach
16	Delemont
17	Porrentruy
18	Altkirch
19	Wittenheim
20	Thann
21	Meyenheim
22	Wolfgantzen
23	Munster
24	SMM
25	Marckolsheim
26	Barr
27	Urmatt
28	Fegersheim
29	Brumath
30	Saverne
31	Soufflenheim
32	Reichshoffen
33	Hoffen
34	Kandel
35	BadBergzabern
36	Dahn

Output data of scenario

Straw ConRec

Biomass utilised:
465 kt/a

Number of facilities:
44

Total distance:
1 893 km

Total transport costs:
6.32 Mio.€

Total biomass provision costs:
164 Mio.€

Total investments:
57 Mio.€

Total energy:
71 MW_{el} (DG/FBG/Comb)
21.4 MW_{th} (Comb)

Figure 4.27: Scenario results of straw ConRec.

The straw potential of 465 kt/a is combusted and gasified for the production of electric and thermal bioenergy. Whereas DG and FBG gasification technologies are used for the generation of 62.4 MW_{el} of electricity, combustion produces additionally 8.6 MW_{el} of electricity and 21.4 MW_{th} of heat energy. The four combustion facilities range from 1.8-2.3 MW_{th} and 4.5-5.8 MW_{el} and are located in Appenweiler (5), Wittenheim (19), Urmatt (27) and Dahn (36). Gasification facilities, especially with a down draft gasifier (DG), are distributed homogeneously in the URR with a constant capacity of 1.75 MW_{el}. FBG technology facilities on the other hand are located at the same four combustion facility locations with capacities ranging between 1.3-1.74 MW_{el}. The resulting network structure is diverse with the four aforementioned main energy generating centres.

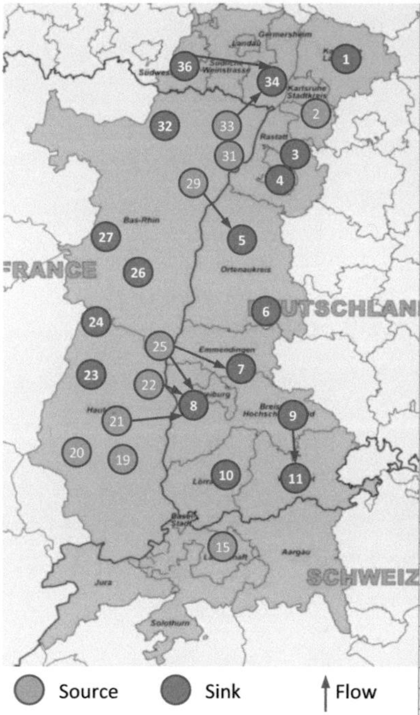

ID	Location
01	Bruchsal
02	Ettlingen
03	Gaggenau
04	Geroldsau
05	Appenweier
06	Haslach
07	Denzlingen
08	BadKrozingen
09	Titisee
10	Zell
11	Haeusern
12	Frick
13	Wohlen
14	Balsthal
15	Sissach
16	Delemont
17	Porrentruy
18	Altkirch
19	Wittenheim
20	Thann
21	Meyenheim
22	Wolfgantzen
23	Munster
24	SMM
25	Marckolsheim
26	Barr
27	Urmatt
28	Fegersheim
29	Brumath
30	Saverne
31	Soufflenheim
32	Reichshoffen
33	Hoffen
34	Kandel
35	BadBergzabern
36	Dahn

Output data of scenario:

Vineyard residues (prunings) ConRec

Biomass utilised:
75.6 kt/a

Number of facilities:
18

Total distance:
387 km

Total transport costs:
0.28 Mio.€

Total biomass provision costs:
23 Mio.€

Total investments:
7.8 Mio.€

Total energy:
15.2 MW_{el} (DG)

Figure 4.28: Scenario results of vineyard residues (prunings) ConRec.

The 18 small-scale gasification facilities converting the 75.6 kt/a of woody vineyard residues into 15.2 MW_{el} of electric bioenergy are sparsely supplied with feedstock from other areas. Regional woody vineyard residues are provided by vineyard cultivation in the surrounding areas. The capacity range of the DG gasifiers varies between 0.1-1.75 MW_{el} with an average of 0.84 MW_{el} and average investments of about 0.4 Mio.€ per facility.

4.2.3.2 Results of the model for the MaxEx scenario

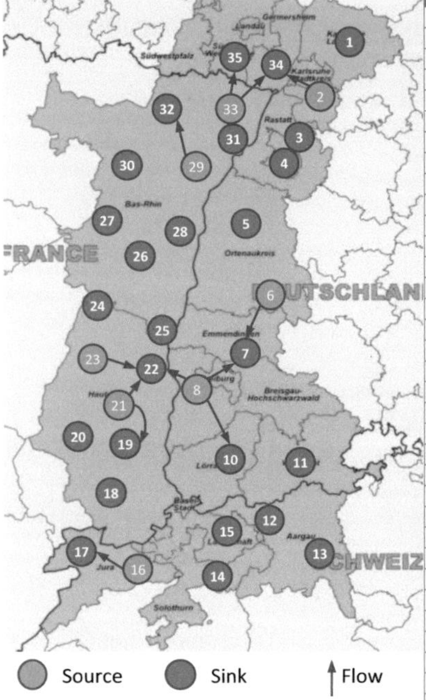

ID	Location
01	Bruchsal
02	Ettlingen
03	Gaggenau
04	Geroldsau
05	Appenweier
06	Haslach
07	Denzlingen
08	BadKrozingen
09	Titisee
10	Zell
11	Haeusern
12	Frick
13	Wohlen
14	Balsthal
15	Sissach
16	Delemont
17	Porrentruy
18	Altkirch
19	Wittenheim
20	Thann
21	Meyenheim
22	Wolfgantzen
23	Munster
24	SMM
25	Marckolsheim
26	Barr
27	Urmatt
28	Fegersheim
29	Brumath
30	Saverne
31	Soufflenheim
32	Reichshoffen
33	Hoffen
34	Kandel
35	BadBergzabern
36	Dahn

Output data of scenario:

Forest residues MaxEx

Biomass utilised: 130.84 kt_{atro}/a

Number of facilities: 26

Total distance: 484 km

Total transport costs: 0.43 Mio.€

Total biomass provision costs: 34 Mio.€

Total investments: 7.6 Mio.€

Total energy: 14.77 MW_{el} (DG)

Figure .4.29: Scenario results of forest residues MaxEx.

The forest residues biomass potential of the MaxEx scenario is assumed to be approximately 130.84 kt_{atro} per year and can be utilised for energy production of around 14.77 MW_{el} via down draft (DG) gasification with a conversion capacity range of 0.11-1.5 MW_{el}. The average energy output is 0.6 MW_{el} with average investments of 0.3 Mio.€ per facility.

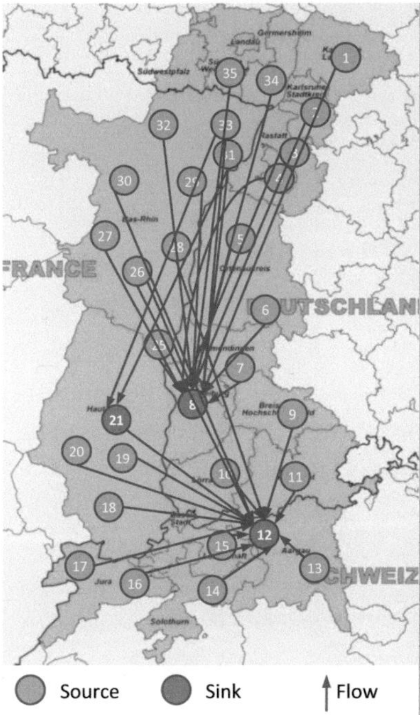

ID	Location
01	Bruchsal
02	Ettlingen
03	Gaggenau
04	Geroldsau
05	Appenweier
06	Haslach
07	Denzlingen
08	BadKrozingen
09	Titisee
10	Zell
11	Haeusern
12	Frick
13	Wohlen
14	Balsthal
15	Sissach
16	Delemont
17	Porrentruy
18	Altkirch
19	Wittenheim
20	Thann
21	Meyenheim
22	Wolfgantzen
23	Munster
24	SMM
25	Marckolsheim
26	Barr
27	Urmatt
28	Fegersheim
29	Brumath
30	Saverne
31	Soufflenheim
32	Reichshoffen
33	Hoffen
34	Kandel
35	BadBergzabern
36	Dahn

Output data of scenario:

Household waste MaxEx

Biomass utilised: 670 kt/a

Number of facilities: 2 (3)

Total distance: 3 408 km

Total transport costs: 6.57 Mio.€

Total biomass provision costs: 23.86 Mio.€

Total investments: 43 Mio.€

Total energy: 56.35 MW_{el} (WtE)

Figure 4.30: Scenario results of household waste MaxEx.

With about 670 kt/a of household waste, the MaxEx scenario results detect two main facility locations for the generation of 56.35 MW_{el} of electric energy (i.e., BadKronzingen and Frick). Both of the facilities reach a capacity of about 28.2 MW_{el} with investments of 21.5 Mio.€ each. The network structure is centralised with a total distance of 3 408 km.

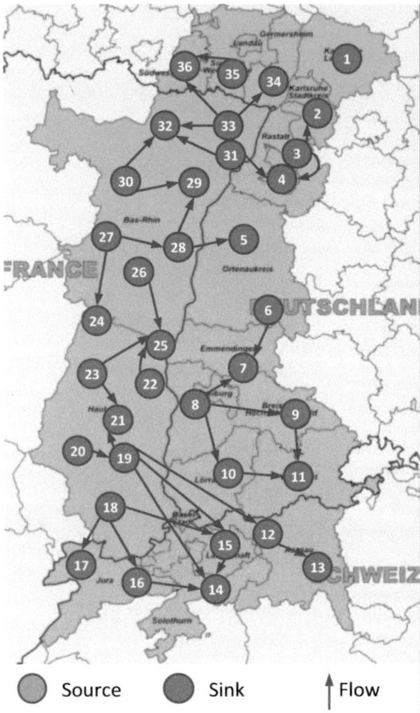

ID	Location
01	Bruchsal
02	Ettlingen
03	Gaggenau
04	Geroldsau
05	Appenweier
06	Haslach
07	Denzlingen
08	BadKrozingen
09	Titisee
10	Zell
11	Haeusern
12	Frick
13	Wohlen
14	Balsthal
15	Sissach
16	Delemont
17	Porrentruy
18	Altkirch
19	Wittenheim
20	Thann
21	Meyenheim
22	Wolfgantzen
23	Munster
24	SMM
25	Marckolsheim
26	Barr
27	Urmatt
28	Fegersheim
29	Brumath
30	Saverne
31	Soufflenheim
32	Reichshoffen
33	Hoffen
34	Kandel
35	BadBergzabern
36	Dahn

Output data of scenario:

Straw MaxEx

Biomass utilised:
744 kt_{atro}/a

Number of facilities:
58

Total distance:
1 469 km

Total transport costs:
7.5 Mio.€

Total biomass provision costs:
263 Mio.€

Total investments:
102 Mio.€

Total energy:
86 MW_{el}
(DG/FBG/Comb)
60.4 MW_{th} (Comb)

Figure 4.31: Scenario results of straw MaxEx.

By comparing the straw ConsRec scenario with the MaxEx scenario, the number of small-scale gasification and combustion facilities increases by 14 to 58 forming a very inhomogeneous structure with 11 FBG (min.: 1.33 MW_{el}, max.: 1.7 MW_{el}, avg.: 1.6 MW_{el}) and 25 1.75 MW_{el} DG gasification as well as 22 combustion (min.: 1.82 MW_{th}, max.: 5.8 MW_{th}, avg.: 3.85 MW_{th}) facilities. The resulting network structure is very decentralised with low-capacity technology applications.

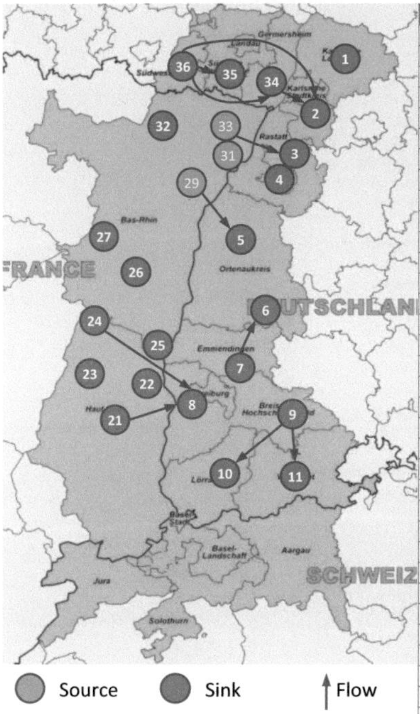

ID	Location
01	Bruchsal
02	Ettlingen
03	Gaggenau
04	Geroldsau
05	Appenweier
06	Haslach
07	Denzlingen
08	BadKrozingen
09	Titisee
10	Zell
11	Haeusern
12	Frick
13	Wohlen
14	Balsthal
15	Sissach
16	Delemont
17	Porrentruy
18	Altkirch
19	Wittenheim
20	Thann
21	Meyenheim
22	Wolfgantzen
23	Munster
24	SMM
25	Marckolsheim
26	Barr
27	Urmatt
28	Fegersheim
29	Brumath
30	Saverne
31	Soufflenheim
32	Reichshoffen
33	Hoffen
34	Kandel
35	BadBergzabern
36	Dahn

Output data of scenario:

Vineyard residues (prunings) MaxEx

Biomass utilised: 122 kt_{atro}/a

Number of facilities: 22

Total distance: 549 km

Total transport costs: 0.52 Mio.€

Total biomass provision costs: 36.8 Mio.€

Total investments: 7.6 Mio.€

Total energy: 24.4 MW_{el} (DG)

Figure 4.32: Scenario results of vineyard residues (prunings) MaxEx.

As the vineyard residues (prunings) potential increases in the MaxEx scenario up to 122 kt/a, the number of DG gasification facilities rises to 22. The capacities range between 0.1 and 1.75 MW_{el} with an average of 1.1 MW_{el}, which is a slight growth of 0.26 MW_{el} in comparison to the ConsRec scenario. In alignment with the previous results of the vineyard prunings, two main utilisation areas in the URR can be characterised, i.e., the North(-East) and the central area of the URR, which contain sufficient vineyard residues to be utilised on-site.

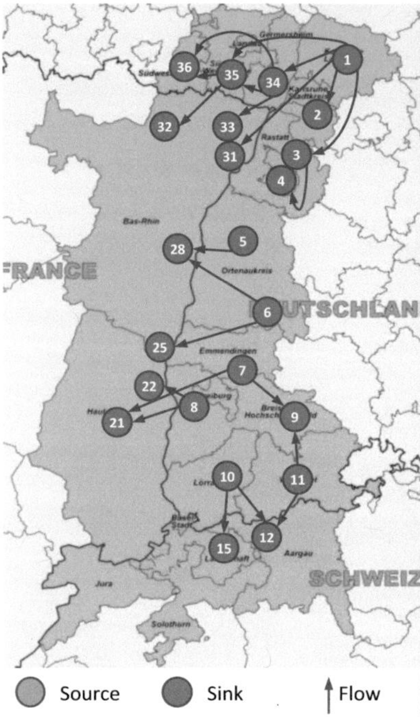

ID	Location
01	Bruchsal
02	Ettlingen
03	Gaggenau
04	Geroldsau
05	Appenweier
06	Haslach
07	Denzlingen
08	BadKrozingen
09	Titisee
10	Zell
11	Haeusern
12	Frick
13	Wohlen
14	Balsthal
15	Sissach
16	Delemont
17	Porrentruy
18	Altkirch
19	Wittenheim
20	Thann
21	Meyenheim
22	Wolfgantzen
23	Munster
24	SMM
25	Marckolsheim
26	Barr
27	Urmatt
28	Fegersheim
29	Brumath
30	Saverne
31	Soufflenheim
32	Reichshoffen
33	Hoffen
34	Kandel
35	BadBergzabern
36	Dahn

Output data of scenario

Woody biomass MaxEx

Biomass utilised: 332 kt/a

Number of facilities: 23

Total distance: 876 km

Total transport costs: 2 Mio.€

Total biomass provision costs: 86.6 Mio.€

Total investments: 19.3 Mio.€

Total energy: 37.5 MW_{el} (DG)

Figure 4.33: Scenario results of woody biomass MaxEx.

The potential of woody biomass of 332 kt/a is only sufficient in the MaxEx scenario and results in 23 small-scale DG gasification facilities generating 37.5 MW_{el} with a constant capacity of 1.75 MW_{el}.

By allowing more facilities at potential sink locations, the number of facilities decreases to 13 with key locations in Bruchsal (1), BadKronzingen (8), and Kandel (34). In general, the network structure is decentral with only little biomass transportation.

4.2.4 Summary of section 4.2

Table 4.12: Summary of selected results of the modelled scenarios.

		Biomass [kt/a]	Facilities [#]	Distance [km]	Transport costs [Mio.€]	Provision costs [Mio.€]	Investments [Mio.€]	Energy output [MW]	Costs per energy [Mio.€/MW]
Forest residues	2010	163	31	336	0.5	43	9.5	18.5	2.865
	BAU	16.869	16	639	0.06	3.88	0.863	1.68	2.859
	ConsRec	81,775	24	619	0.3	21	4.7	9.23	2.817
	MaxEx	130,84	26	484	0.43	34	7.6	14.77	2.846
House Hold waste	2010	841	3	2343	5.9	30	54	70.7	1.272
	ConsRec	335	3	2139	3.26	12	21.5	28.2	1.304
	MaxEx	670	1	3408	6.57	23.86	43	56.35	1.303
Straw	2010	930	3	1522	9.5	330	133	182.1	2.595
	BAU	93	66	431	0.45	33	10	18.7	2.324
	ConsRec	465	4431	1893	6.32	164	57	92.4	2.460
	MaxEx	744	58	1469	7.5	263	102	146.4	2.544
Vin. res. prunings	2010	151	24	576	0.7	46	16	30.6	2.056
	BAU	14,3	11	370	0.053	4.3	1.5	2.86	2.047
	ConsRec	75,6	18	387	0.28	23	7.8	15.2	2.045
	MaxEx	122	22	549	0.52	36.8	7.6	24.4	1.841

The model results of the main four types of biomass feedstock conclude that an increased potential does not necessary lead to higher costs per energy. The choice of technology, capacity, and the number of facilities as well as the structure of the logistics network is crucial. By comparing the costs per energy, the energetic utilisation of household waste seems to be the most economic followed by vineyard residues, straw, and forest residues.

4.2.5 References section 4.2

Bidart, C, McKenna, R, Fichtner, W 2014-2015, Research Reports & Protocols of the 'OUI Biomasse' Research Area 3, Karlsruhe.

Kerdoncuff, P 2008, *Modellierung und Bewertung von Prozessketten zur Herstellung von Biokraftstoffen der zweiten Generation*, KIT Dissertation, Karlsruhe.

Rosenthal, RE 2012, GAMS - A User's Guide, Available from: <http://www.gams.com/dd/docs/bigdocs/ GAMSUsersGuide.pdf> [13 August 2012].

Schwaderer, F 2012, *Integrierte Standort-, Kapazitäts- und Technologieplanung von Wertschöpfungsnetzwerken zur stofflichen und energetischen Biomassenutzung*, KIT Dissertation, Karlsruhe.

4.3 Environmental impacts of bioenergy production and utilisation and sustainability assessment of the scenarios

Nadège Blond[1], *Mathieu Bosansky*[2], *Joëlle Duplay*[3], *Yaxian Hu*[4], *Martin Knapp*[5], *Rolf Meyer*[5], *Alexandra Pehle*[5], *Cyril Pallarès*[2], *Jean-Luc Ponche*[1,6], *Gaëtana Quaranta*[3], *Wissal Selmi*[1], *Gerhard Schäfer*[3], *Jérémy Stark*[1], *Volker Stelzer*[5]

[1] Laboratoire Image, Ville, Environnement (LIVE), Université de Strasbourg/ CNRS, Strasbourg, France.

[2] Association pour la Surveillance de la Pollution Atmosphérique (ASPA), Strasbourg, France.

[3] Laboratoire d'HYdrologie et de GEochimie de Strasbourg (LHyGeS), Université de Strasbourg/ CNRS, Strasbourg, France.

[4] Departement Umweltwissenschaften (DUW), Universität Basel, Basel, Schwitzerland.

[5] Institute for Technology Assessment and Systems Analysis (ITAS), Karlsruhe Institute of Technology (KIT), Karlsruhe, Germany.

[6] Institut de Chimie et Procédés pour l'Energie, l'Environnement et la Santé (ICPEES), Université de Strasbourg/CNRS, Strasbourg, France.

Bioenergy, as any other type of energy, is produced through the transformation of a system from which it is extracted. As evidence, the system is impacted and the impacts may occur at the different steps of the transformation of that system.

Indeed, when biomass is produced, several impacts can occur. The land use changes with possible land-use conflicts. Soil contents and the soil-air exchanges are modified by the growth or removal of plants and this even more when fertilizers are used in cultivation. The use of machineries from the sowing until the harvest and the transport of biomass modifies the air-soil fluxes, consumes fuels (mainly fossil fuels) with emissions of chemical compounds in the environment. When the biomass is transformed into bioenergy, air pollutants are also usually emitted: some of them are directly harmful to local ecosystems; and others with longer residence time in the atmosphere and global warming potential are contributing to global changes.

Nevertheless bioenergy is expected to be produced in a sustainable way from local renewable resources. The growth of biomass contributes to absorb greenhouse gases and stock elements in soils. The conversion of biomass into bioenergy can be controlled to minimise as much as possible air pollutant emissions and impacts. All steps of industrial processes (including extraction, transport, etc.) from biomass to bioenergy should be assessed with the aim to reduce impacts as far as possible.

The overall objective of a sustainable use of biomass is to control and reduce pressures on resources, dependency on fossil energy reserves and anthropogenic impacts. Creation of a local economy and associated employments are also foreseen.

This section aims to present the results of the Research Area (RA) five of the 'OUI Biomasse' project. The objective of this RA was the evaluation of the impacts and more generally the sustainability of several pathways of biomass and bioenergy production and utilisation over the Upper Rhine Region. The work was divided in several tasks:

- A modelling work was performed to evaluate the sinks and sources of greenhouse gases (GHG) due to the so-called "landuse, landuse change and forestry" (LULUCF) activity sector and assess the potential impacts of an increase production of biomass for bioenergy production;
- Since Miscanthus is seen as a promising bioenergy crop with possible adding soil carbon sequestration, few Miscanthus fields were sampled to study their impacts on soil organic content, soil respiration and PH;
- Life Cycle Assessment was applied to assess the impacts of several types of biomass (and mixtures), that are wood chips, wheat, grape marc, and Miscanthus;
- The RA4 scenarios were detailed to assess the impacts of an increasing bioenergy production and consumption on the air emissions and associated air pollution.

- Since the increasing utilisation of biomass also brings along great challenges in sustainability due to social, ecological and economic impacts (Upreti, 2004), a task was finally designed to give an overview of the sustainability of the RA4 scenarios.

This work sparks a discussion on how the expanding bioenergy sector interacts with several society demands relative to the protection of the environment (air, soil and nature conservation, biodiversity), the food production, and more generally the economy.

4.3.1 Impacts of biomass production on soil-air exchanges

***Jérémy Stark*[1], *Nadège Blond*[1], *Mathieu Bosansky*[2], *Wissal Selmi*[1], *Jean-Luc Ponche*[1,3]**

[1] Laboratoroire Image, Ville, Environnement (LIVE), Université de Strasbourg/CNRS, Strasbourg, France;

[2] Association pour la Surveillance de la Pollution Atmosphérique (ASPA), Strasbourg, France;

[3] Institut de Chimie et Procédés pour l'Energie, l'Environnement et la Santé (ICPEES), Université de Strasbourg/CNRS, Strasbourg, France.

In Europe, the land is already widely exploited to produce food (vegetal and animal), raw materials for industry (furnishing, construction, etc), and biomass to be converted in bioenergy. An overview of the URR situation was given in section 2. These human practices, and especially when fertilizers are used, modify the air pollutant exchanges between the soil, the plant and the atmosphere.

The objective of this section is to evaluate the potential impact of an increase exploitation of the soil for bioenergy production on the greenhouse gases (GHG) air-soil exchanges. The study focuses on the LULUCF activity sector (Land-Use, Land-Use Change and Forestry) that is part of the classification of the air pollutant activities. It repertories sinks and sources of greenhouse gases (GHG) due to the soils and the forests, and excludes all the activities linked to direct human pressures, like use of machineries and fertilizers. These last activities are already classified in other activity sectors like transportation, agriculture, etc. and will be studied in the section 0. Bosanski (2010) proposed a first GHG budget of the LULUCF sector for the Alsace Region. Based on quiet similar methodologies, the aim is here to analyse this budget at the URR scale.

4.3.1.1 GHG sinks and sources due to growth of forest and wood harvest

Forestry is an important sink of carbon that is mainly determined by the type and growth of the forest, while the harvest of the wood leads to an emission of CO_2. The methodology to compute the net carbon balance for forestry is based on the use of statistics on the quantity of merchant wood, i.e. stem wood from trees with a diameter larger than 7.5 cm at 130 cm above ground. These data were collected for France from National Forest Inventory for the year 2010 (IFN 2010), for Germany from the Bundeswaldinventur for 2002, for Switzerland, from the Landesforestinventar for 2006. Note that the inventories do not concern the same year but were the most recent available data for the study. For France, data on the harvest per tree species was also given by (EAB 2012) and completed from data issued from (FIBOIS 2010). The resulting yearly carbon balance “C” can be written as the combination of several terms:

- the quantity of merchant stem wood (MW, in m^3),
- an expansion factor (BEF, no unit) that allows to estimate the biomass issued from the rest of the tree (branches, roots, leaves),
- the biomass density (D, in tons of dry matter per m^3, tDM/m^3) that is the quantity of dry matter per quantity of biomass
- the carbon fraction content (CF, tC/tDM), i.e. the carbon emission factor given per ton of dry matter.

The C balance is then calculated as followed:

$$C = MW_G * BEF_G * D * CF - MW_H * BEF_H * D * CF$$

where MW_G and MW_H respectively correspond to growing and harvested merchant wood, BEF_G and BEF_H are the corresponding expansion factors (BEF_G is used to compute the total growing biomass including the roots; BEF_H excludes the root system that is not harvested). The release of carbon due to harvest (second term) is assumed to occur during the year of the harvest. Such assumption is valid for short life cycle products like fuel wood, paper or cardboard but can be limited for long life cycle products and may lead to an overestimation of the carbon emissions.

Among literature, the most recent values of 1.25 and 1.6 were chosen for respectively BEF_H and BEF_G (CITEPA 2012). The wood densities were extracted from (Diets 1975): values are respectively of 0.5 and 0.41 tDM/m^3 for deciduous and coniferous trees. Finally carbon contents were also studied on a literature basis (Tobin and Nieuwenhuis 2005; Löwe et al. 2002; INRA 2002; AGRIGES 1999) the CF mean value of 0.475 tC/tDM was selected. It is assumed that all this carbon is exchanged through CO_2 molecules (CO_2 balance is computed from carbon balance C using the mass ratio 3.67tCO2/tC). The C balance of the forestry leads to a global URR sink of -5'545 kTeq CO_2 per year due to the forest growth.

4.3.1.2 GHG sinks and sources due to landuse change

The land-use and the land-use changes may also lead to other sources and sinks of GHG. They were analysed on the basis of the changes between 1990 and 2006 using CORINE LAND COVER (CLC) database (EEA 2009) for these two years. Five categories of land-uses were extracted: wetland, forest, grassland, cropland and urban area. Table 4.13 gives the surfaces affected by land-use changes on the URR between 1990 and 2006. The most important changes concern forests-grasslands. The net balance indicates that more forests have been converted into grassland than grasslands into forests. The second most important changes have been the transformation of croplands into both urban area and grasslands. The changes finally reach around 2% of the total URR surface between 1990 and 2006.

Note that a large forest area was destructed during the storm Lothar in 1999 (11 900 ha was destructed at 70% in the Alsace Region; FIBOIS 2010).

Table 4.13: Surfaces of land-use and land use changes between 1990 and 2006 (in ha). Grey boxes indicate the non-modified surfaces.

		Land-Use in 2006 (ha)			
		Forest	Grass-land	Crop-land	Urban area
Land-Use in 1990 (ha)	Forest	894 765	23 421	75	571
	Grass-land	9 609	153 342	366	981
	Crop-land	36	2 443	800 736	9 750
	Urban area	15	83	176	193 691

The carbon stock for each of the land-use categories was computed based on measurement data collected over the URR (for France from ARAA, Association pour la Relance Agronomique en Alsace, for Germany from LGRB, Landesamt für Geologie, Rohstoffe und Bergbau, for Switzerland from ETH, Institut für Terrestrische Ökosysteme Boden und Terrestrische Umweltphysik).

These data include measures in different soil depths (E_l in m) of the organic carbon content (C_l in gC/g), the bulk density (D_l in g/m^3), and the coarse fraction (%).The carbon stock S_l (in g/m^2) in the depth 0-30cm was computed for each type of soil as followed (Schwartz and Namri 2002):

$$S_l = C_l.D_l.E_l(1 - G_l)$$

In a second step, the carbon stocks of the five land-use categories were computed as the mean values of the categories and associated to uncertainties derived from the standard deviations. Table 4.14 gives the estimations based on local measurements, compared with other studies or dataset (INRA 2002; data collected from JRC (Hiderer 2013)). It is noted that forest and grassland soils are supposed to contain more carbon than cropland soils. One can note also the high variability in one soil category and the general consistency between data obtained from several sources.

Table 4.14: Carbon stocks and standard deviations (STD) (in tC/ha) per land-use category. Computations based on data collected through several sources. Urban values that were computed as half of the maximum of other carbon stocks (CITEPA, 2012).

		Carbon stocks (STD)			
Source	Region	Forest	Grassland	Cropland	Urban area
Local data used					
ARAA	FR	-	83 (44)	60 (49)	41
LGRB	GR	64 (46)	64 (43)	47 (24)	32
ETH	SW	83 (18)	-	-	42
Other studies					
JRC (Hiderer 2013)	FR	56 (35)	54 (30)	46 (34)	28
	GR	62 (26)	64 (25)	48 (26)	31
	SW	63 (33)	74 (34)	55 (40)	37
INRA, 2002	FR	70	70	43	35

A carbon emission factor ($EF_{l1\rightarrow l2}$ $in\ gC/ha/year$) associated to a land-use change (from l_1 to l_2) as defined as the differences between the initial and the final expected stocks of the land-use categories (respectively S_{l1} and S_{l2}) as followed:

$$EF_{l1\rightarrow l2} = \frac{S_{l2} - S_{l1}}{\Delta t}$$

where Δt (in year) is the duration of the release or storage, fixed here to 20 years.

Since in forests carbon can also be stored in litter and deadwood, or released in case of a land-use change, the same methodology is used with initial or final carbon stocks set to zero. Following a literature overview, the carbon stocks have been chosen to 4.1tC/ha for deadwood (IFN 2010) and 7.1tC/ha for litter (Dupouey et al. 2000). In these cases, the storage is considered as a long term process (Δt=20 years), and the release as a short term process (Δt=1 year only) (CITEPA 2012).

Finally, by considering the emission factors estimated from local measurements and the yearly areas of land-use changes (a sixteenth of the change noted between 1990 and 2006 was assigned for each year), the yearly carbon sinks and sources were computed. It is assumed that all this carbon is emitted as CO_2.

When natural soils (grass-land and forest) are converted into croplands, the soil also releases some nitrous oxide N_2O in the atmosphere (denitration process). The N_2O emissions are linked to the CO_2 emissions with a ratio of $2.7.10^{-3}$tN2O/tCO_2 (IPCC 2003).

The emissions are expressed in tons equivalent CO_2 (teqCO_2) taking into account the N_2O global-warming potential, that is 310 time higher than CO_2 ones (IPCC 2003).

The landuse changes finally leads to emissions of 126kteq CO_2, that is quite small compared to the sink due to forest growth.

4.3.1.3 CH4 balance due to landuse

Methane (CH_4) can be produced or destructed in the soil depending on the hydrological conditions: Methane is produced over flooded and wet soils through digestion of the organic matter by bacteria while CH_4 is captured by methanotrophic bacteria in non-water (dry) soils. CH_4 emissions factors were issued from Roger and Le Mer (2003). These are given for several soils: 433gCH_4/ha/day for wetlands, -5.5gCH_4/ha/day for cultivated soils, -6.5gCH_4/ha/day for grasslands, -9.9gCH_4/ha/day for forest soils (data comparisons were done based on literature but results are not shown here). The balance is expressed for year 2006 in tons equivalent CO_2 (teqCO_2) taking into account the CH_4 global-warming potential, that is 21 time higher than CO_2 (IPCC 2003). A sink of -156kteq CO_2 has been computed, that is also quite small compared to the sink due to forest growth.

4.3.1.4 GHG sinks and sources due to LUCUCF and possible scenarios

Finally, Table 4.15 gives the total GHG balance evaluated for the whole URR for the LULUCF activity sector.

Table 4.15: GHG balance of the LULUCF activity sector over the URR (in kteq CO_2/year). Negative values indicate a sink, positive values a source of GHG.

Subsector	GHG sources and sinks (kteq CO_2/year)	Contribution in percentage
Forestry	-5 545	99%
Land-use change	126	2%
Land-use	-159	3%
Total LULUCF	-5 578	100%

It is noted that the forestry subsector is the driver and consists in an important sink of GHG mainly due to the growth of the forest. The global LULUCF sink corresponds to 17% of the total CO_2 emitted by all activity sectors (estimated to around 34 300kt/year according to (TRION 2012) and data from ASPA).

The uncertainties on the forestry sector was estimated to around 15%, while the ones on others subsectors are much larger because of the high variabilities in land-uses that cannot be enough modelled in the methodology. According to the 'OUI Biomasse' project scenarios (that state that the land use changes in all three scenarios concern arable lands, modified into urban area, and in the MaxEx scenario permanent grassland going to arable lands; the forest area remains unchanged), it is expected that the net GHG budget should not be significantly impacted.

4.3.2 Impact of the production of Miscanthus on the quality of the soils and soil-air exchanges

***Yaxian Hu*[1], *Joëlle Duplay*[2], *Gerhard Schäfer*[2]**

[1] Departement Umweltwissenschaften (DUW), Universität Basel, Basel, Schwitzerland;

[2] Laboratoire d'HYdrologie et de GEochimie de Strasbourg (LHyGeS), Université de Strasbourg/ CNRS, Strasbourg, France;

Miscanthus is seen as a very promising bioenergy crop (Rowe et al. 2009) as its cultivation allows for environmental benefits such as high water use efficiency, no requirement of additional nutrients, and year round cover to reduce soil erosion risk. While most of the research also claimed positive benefits such as mitigating climate change with additional organic carbon sequestration (Howlett et al., 2013; Poeplau and Don, 2014), it is also said that greenhouse gas emissions (GHG) during land use changes might render the net environmental benefits less favorable, especially when large-scale conversion occurs (van der Hilst et al., 2012; Roth et al., 2013). Supplementary negative environmental impacts could be further added up by GHG emissions from applying nitrogen fertilizer, fossil energy consumption during tillage operation, production, storage, transportation and pelletizing (Smeets et al., 2009; Yang and Zhang, 2011; Murphy et al., 2013). In addition, long-term Miscanthus cultivation may also affect soil structure and quality (Christian et al., 1997).

A close investigation onto the quality and mineralization potential of Miscanthus-derived SOC is needed before to promote enhanced plantation of Miscanthus. Thus in complement to the evaluation of GHG balance in the LULUCF sector in the URR, measurements were performed over three soil types from three Miscanthus fields over different cultivation years.

It aims to: 1) detect the potential of Miscanthus in sequestrating organic carbon in the soil; 2) improve our understanding of environmental impacts of long-term and large-scale Miscanthus cultivation.

4.3.2.1 Methodologies for soil sampling and laboratory analysis

Three silty loams collected from four fields were investigated in this study: 1) a silty loam was sampled in May, 2014 from the Farm Niedererweiher, near Ammertzwiller, Alsace, France, after planting 5 years and 20 years of Miscanthus (hereafter termed as "A-5yr", and "A-20yr"). An undisturbed grassland (hereafter termed as "A-grass"), about 50 m away from the 20yr Miscanthus, was also sampled as reference site. 2) A second silty loam was sampled in November, 2014 from the Farm Untergruth in Münchenstein, Basel-Land, from a field after planting 20 years of Miscanthus (hereafter termed as "M-20yr"). A grassland right beside the Miscanthus field was also sampled as reference site (hereafter termed as "M-grass"). 3) A third silty loam was sampled in November, 2014 from the Farm Hofgut Farnsburg in Ormalingen, Basel-Land, Switzerland after planting 20 years of Miscanthus (hereafter termed as "F-20yr"). A grassland right beside the Miscanthus field was also sampled as reference site (hereafter termed as "F-grass"). In general, after harvesting in March or April, about 30 cm high Miscanthus residues are left standing. No fertilizer of any kind has ever been applied to the either Miscanthus fields or grassland. For the Farm Niedererweiher (Alsace), 500kg.ha^{-1} agricultural lime is applied every 5 years to neutralise the soil acidity. No agricultural lime has been applied to the other two fields at Münchenstein and Farnsburg. Soil cores from four layers, 0-10cm, 10-40cm, 40-70cm and 70-100cm, were collected separately on all the grasslands and Miscanthus fields. The sampling spots on each field were randomly chosen, and at least three repetitions were carried out. Surface soils were also collected by cylinders to determine soil bulk density.

Respiration rates were measured based on the method described in (Robertson et al., 1999) and (Zibilske, 1994). About 25g of (moist) soils from top 10 cm and 10-40 cm from all three fields were immediately incubated at 20°C in flasks with volume of 200 cm^3 (flasks open). Visible residues or roots were manually removed as much as possible. Prior to soil respiration measurements, all flasks were sealed using rubber stoppers. Differences in CO_2 concentrations between the 1h period of time were used to calculate the instantaneous respiration rate. The respiration rate measurements were repeated at day 1, 2, 3, 7, 14, and every 7 days since then (42 days in total). The CO_2 concentrations were measured using a SRI8610C Gas Chromatograph. During the incubation period, the wet weight of each soil sample was monitored every 3 days, and the variation of soil moisture was constrained within 1%.

The stable isotopes of $\delta^{13}C$ and $\delta^{15}N$ of all the layers on each sampling site were analysed using an Isotope Ratio Mass Spectrometry at University of California, Merced to determine if and how much Miscanthus contributes to changing SOC compounds. The stable isotopic compositions were expressed in δ notation (‰) as: $\delta^{13}C$ or $\delta^{15}N$ = [(Rsample - Rstandard)/ Rstandard] × 1000; where Rsample is the ratio of the heavy to the light C (13C/12C) or N (15N/14N) isotopes in a sample; and Rstandard is the ratio in a standard (Dawson & Brooks, 2001). The standards were referenced to international standards Pee Dee Belemnite. Soil total organic carbon (SOC) contents of all layers were measured using Leco RC612. Ratio of C:N was determined by Leco CN628. The pH values were determined in 0.01M CaCl2 suspension (1:2.5) using a SevenExcellence pH meter. All the data analysis was carried out by R Studio software packages.

4.3.2.2 Analyses of investigated miscanthus fields

Soil organic carbon stock

The distributions of soil organic carbon (SOC) content of all depths from all the fields are shown in Figure 4.34. For all the layers, the SOC content of Miscanthus fields was generally higher than that of the grassland. This is consistent with the results of Hansen et al. (2004), who reported that SOC contents were higher at all soil depths under the 16 year old Miscanthus when compared to reference sites. Yet, in this study, the SOC content in general decreased with soil depths (Figure 4.34), indicating the superiority of Miscanthus in increasing SOC stock. It is more significant in the surface soil and the top 0-10 cm (Figure 4.36). Similar decreasing tendency of SOC content over soil depths was already observed by Zimmermann et al. (2013) and Dufossé et al. (2014). The results confirm their explanation that the pronouncedly greater SOC content in the top soil of the Miscanthus fields is due in part to the return of residues (about 30cm high in our study) helping to cumulate the SOC content on soil surface. In addition, the absence of tillage operation cannot bury surface soils into deeper layers.

Stable isotope $\delta^{13}C$ and $\delta^{15}N$

The $\delta^{13}C$ of the soils from Miscanthus fields was evidently less depleted than that of the soils from the grassland (Figure 4.35). It means that C4 plant Miscanthus has significantly changed the SOC compounds. Such changes are more significant in upper 10-40cm, and gradually diminished through soil depth, implying that the effects of Miscanthus to soil properties were limited in upper layers. The less depleted $\delta^{13}C$ signature in the upper layers than in the lower layers (Figure 4.35) is in line with the more enriched SOC content in the upper layers (Figure 4.34), again suggesting that the effects of Miscanthus to increase SOC stock are limited to the upper layers. The $\delta^{15}N$ (Figure 4.35) showed similar pattern to $\delta^{13}C$ with less depletion in the Miscanthus soils than in the Grassland soils, implying a greater decomposition potential in the Miscanthus soils.

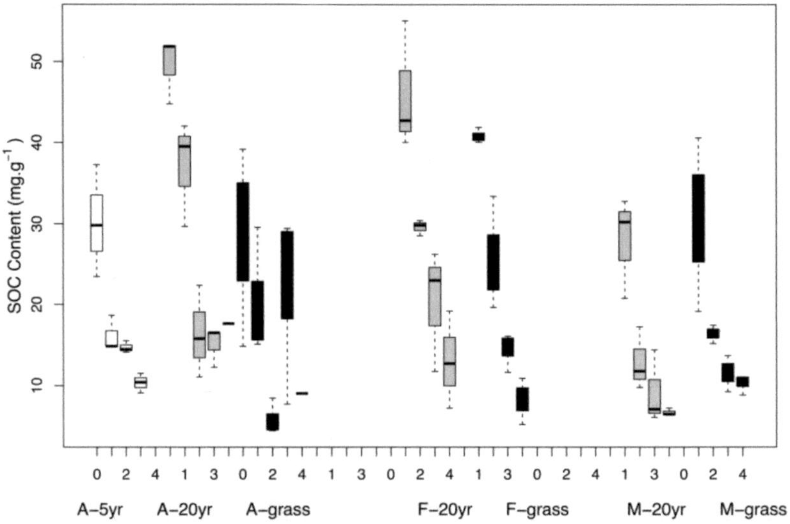

Figure 4.34: Distribution of total organic carbon (SOC) content across different soil depths of the Miscanthus fields and grassland from the three sites in Alsace (A), Farnsburg (F) and Münchenstein (M). The numbers 0, 1, 2, 3, and 4 on the X-axis represent soil layers: Surface, 0-10cm, 10-40cm, 40-70cm and 70-100cm (n=3).

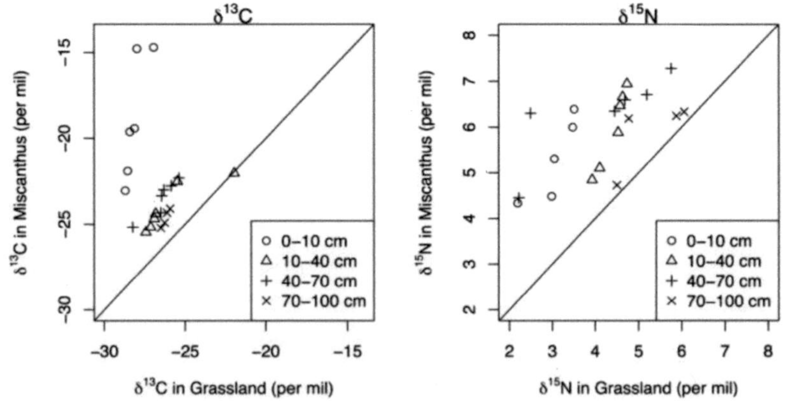

Figure 4.35: Comparison of the $\delta^{13}C$ and $\delta^{15}N$ between Miscanthus fields and grassland across all the layers collected from all the three sites in Alsace, Farnsburg and Münchenstein. The bold line indicates the 1:1 ratio.

4.3.2.3 C: N ratio

In the upper soil layers, the C:N ratios of the Miscanthus soils were evidently greater than that of the grassland soils (Figure 4.36), implying the existence of a potentially great amount of undecomposed residues on the soil surface. This is in good line with the less depleted $\delta^{15}N$ signature in the Miscanthus field illustrated in Figure 4.35. Those undecomposed residues contribute, to an un-known extent, to the greater SOC content in the top soils of the two Miscanthus fields. On one hand, it potentially leads to an overestimation on the net SOC sequestration rates of Miscanthus cultivation; on the other hand, the greater amount of undecomposed residues also promises a greater mineralization potential, and thus additional CO_2 emissions than the grassland soils. The low C:N ratios in deep layers of the two Miscanthus fields are probably caused by the absence of fertilizer and lack of tillage to bring undecomposed residues to deep soils. The C:N ratios observed in this study were lower than that reported in Dufossé et al. (2014), where the C:N ratios only slightly declined with soil depth.

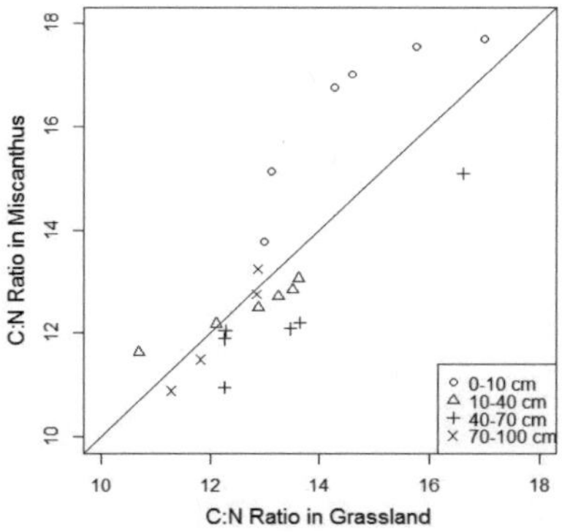

Figure 4.36: Comparison of the C:N ratio between Miscanthus fields and grassland across all the layers collected from all the three sites in Alsace, Farnsburg and Münchenstein. The bold line indicates the 1:1 ratio.

4.3.2.4 CO_2 emissions

The respiration rates of the Grassland and Miscanthus soils of the three sites were pairwise plotted in Figure 4.37. It shows that the respiration rates of Miscanthus in general were slightly greater than that in grassland in both soil layers. The greater respiration rates on the two Miscanthus fields may partially come from the decomposition of unavoidable mixture of roots or residues accumulated on the soil surface, and in another part due to their greater susceptibility of SOC to mineralization. The greater SOC stock in the Miscanthus fields (Figure 4.34) with greater respiration rate (Figure 4.37) may have a great environmental impact, when converting Miscanthus fields into other crop fields. By then, the positive effects of Miscanthus fields in sequestrating SOC ceases, while the labile fraction of SOC may contribute to additional atmospheric CO_2 emissions. This further urges the necessity to determine the quality of SOC so as to fully understand the environmental impacts of Miscanthus cultivation on a wider range of soils.

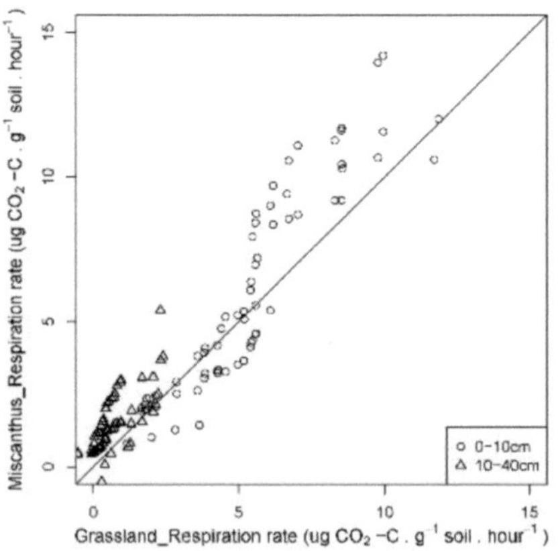

Figure 4.37: Pairwise scatterplots of the respiration rates per gram soil per hour between the Grassland and the Miscanthus fields for both the 0-10cm and 10-40cm layers of the three sites. The "CO_2-C" denotes the net C efflux as CO_2 emissions. The bold line indicates 1:1 ratio.

4.3.2.5 pH

The pH values of all the four soil depths on the three fields are illustrated in Figure 4.38. In general, the Miscanthus fields had lower pH values than the grassland fields, indicating a risk of soil acidification on the Miscanthus fields. Apart from the generally lower pH on the Miscanthus fields, the pH on the top 10 cm of the Miscanthus field was particularly low. While the pH values observed in this study were still within the tolerance limits of soils, the decreasing pH values over long-term Miscanthus cultivation, especially in the top soil, suggest that Miscanthus cultivation bears the risk to cause soil acidification, despite the compensation of agricultural lime applied by the farmer every 5 years. Similar pattern has been observed by Foereid et al. (2004) on Miscanthus fields with different ages, further emphasizing the long-term impacts of Miscanthus onto soil quality.

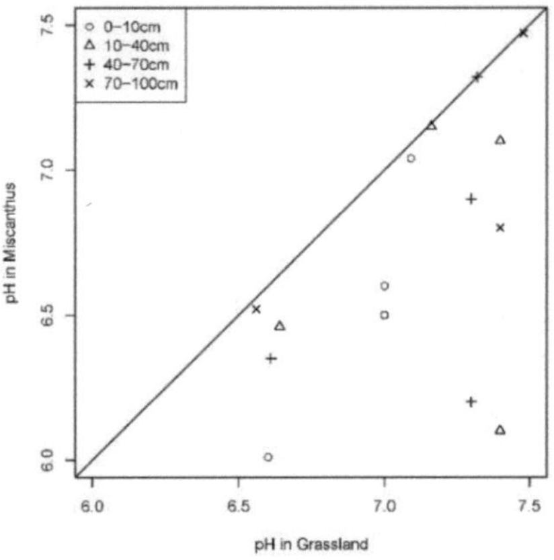

Figure 4.38: Pairwise scatterplots of the pH values between the Grassland and the Miscanthus fields in different layers. The bold line indicates 1:1 ratio.

To summarise, the results show that Miscanthus cultivation can signify-cantly increase the SOC stocks compared to grassland (of around 30% in some cases). However, the benefits of SOC sequestration may only occur on the surface soil, mostly due to the accumulation of residues on the soil surface and the absence of tillage operation to bring the undecomposed residues to deep layers. The results, therefore, caution the use of total SOC stocks on the surface soil to estimate the net benefits of Miscanthus cultivation in terms of GHG mitigation. The potential error to overestimate the benefits of Miscanthus in mitigating climate change further increased with the greater respiration rates of soils from the Miscanthus fields. The risk of acidification adds another precaution to the environmental impacts of Miscanthus cultivation in the entire Upper Rhine Region. More investigations must be conducted on a wider range of soils over various cultivation ages in the future research.

4.3.3 Life Cycle Assessment methodology applied to biomass life cycle

Gaëtana Quaranta

Laboratoire d'HYdrologie et de GEochimie de Strasbourg (LHyGeS), Université de Strasbourg/ CNRS, Strasbourg, France.

Life Cycle Assessment (LCA) aims at addressing environmental problems of a product over its life cycle from cradle to grave. Few studies using LCA (Kaltschmitt et al., 1997; Sai Liang et al., 2013) showed that bioenergy offers some ecological advantages such as conserving fossil energy resources or reducing the greenhouse effect but they also have some definite disadvantages regarding land use and certain airborne pollutant emissions when the overall life cycle is considered. Indeed, biomass production may put pressure on agricultural fields dedicated to food production, and the biomass combustion leads to emissions of harmful pollutants, like particles, nitrogen dioxides, and HAPs. Specific LCA studies compare biomass feedstocks (Butnar et al., 2010; Godard et al., 2012) and demonstrated that Miscanthus had the lowest impact per unit of energy produced, whereas wheat had up to twice-higher impacts due to its high input requirements. These differences were mainly due to discrepancies in yield potentials and fertilizer rates (ECOBIOM Project, 2009). Kim and Dale (2005) showed that biomass cropping systems offer benefits in terms of nonrenewable energy and reduce greenhouse gases, but utilisation of biomass for biofuels also tend to increase acidification and eutrophication, because releases of nitrogen and phosphorus from soil during cultivation.

Using LCA with the standard tool/method (ISO14040, 14044-2006), four different kinds of local biomass and five conversion pathways were studied in the 'OUI Biomasse' project: wood chips, mixture of Miscanthus and wood pellets with grape marc, wheat ethanol. Those biomasses are used for energy production in the URR region for heating or transport. Eleven impact categories were assessed such as resources depletion, ozone depletion, global warming, photo-oxidant formation, carcinogenic effects, respiratory effects, eutrophication, ecotoxicity, etc. These impacts contribute into three damage categories: human health, ecosystem quality and climate change.

The LCA was first applied to a district-heat production and involved wood chips production, transport and wood chips combustion. The system is located in Colmar, Alsace, France (collaboration of the Société Colmarienne de Chauffage Urbain). It was shown that the wood chips production step mainly contributes to the abiotic depletion and global warming due to the petrol consumption and associated CO_2 emissions. The wood chips combustion step mainly contributes to photo-oxidant formation through the emissions of SO_2, CO and CH_4 while its impact on human toxicity is due to the emissions of As, HAP, and NOx. Nevertheless, it was noticed that the combustion emissions are 60% lower than regulatory thresholds (Fig.4.39).

Impact categories	Chips production	Transport, forest-storage	Transport, storage-SCCU	Chips combustion
Abiotic depletion (kg Sb eq)	1.76E-05	6.45E-06	5.02E-07	5.70E-06
Global warming, 100 yrs (kg CO_2 eq)	2.65E-03	8.96E-04	6.98528E-05	1.82E-03
Ozone layer depletion (kg CFC-11 eq)	1.45E-10	1.45E-10	1.13E-11	4.21E-11
Photo-oxidant formation (kg C_2H_4 eq)	7.70E-07	1.72E-07	1.34E-08	1.72E-06
Human toxicity, 100 yrs (kg 1.4 DCB eq)	1.68E-03	1.36E-04	1.06E-05	5.93E-02

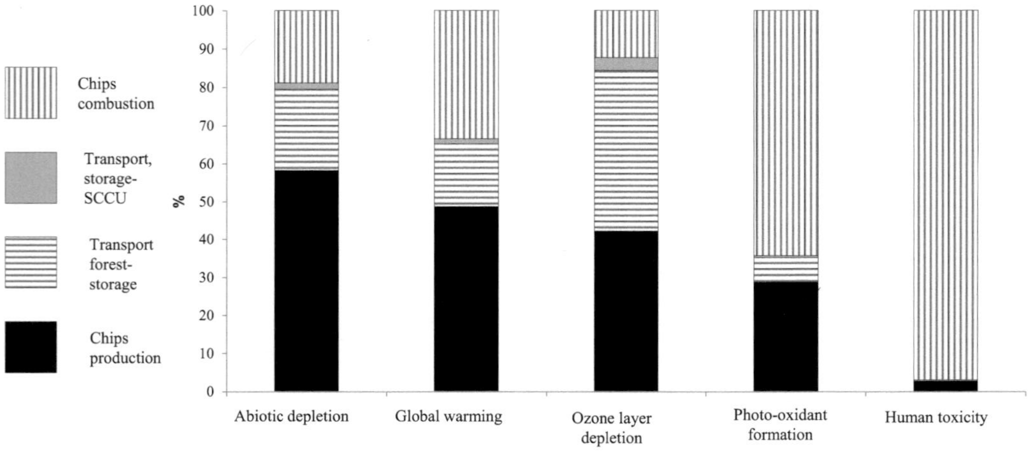

Figure 4.39: District-heat production LCA results.

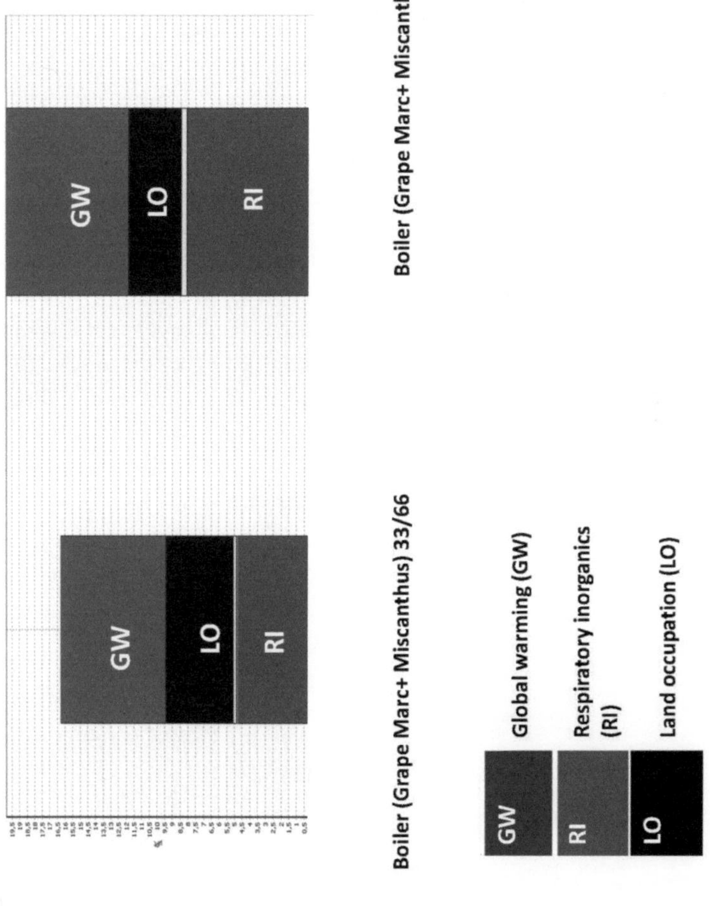

Figure 4.40: Combustion prototype boiler LCA results.

LCA was also applied to the combustion of several biomasses in a prototype boiler (Boiler REKA, type HKRST/V-FSK 20): mixtures of Miscanthus and grape marc (with proportions of 50/50 and 33/66), mixtures of wood pellet and grape marc (Fig.4.40). A compared analysis of the different combustion processes was conducted to understand and assess the various environmental impacts and to get an idea of these different types of fuels. The LCA of the mixtures of Miscanthus and grape marc combustion processes focusses on three impact categories: global warming (linked to CO_2, CH_4 emissions during combustion), inorganic respiratory (linked to NO, CH_4 emissions during combustion) and land occupation. The 50/50 mixing presents more impacts than 33/66 mixing showing that the grape marc is the responsible component of the environmental impact. One can note the same analysis for the mixture of pellet and grape marc.

Finally, a LCA of wheat ethanol was conducted from wheat grain production to ethanol production and combustion phase in a car considering five mixtures of fuels. Only the results obtained for the ethanol production is showed below (Fig. 4.41). It showed a high contribution of the wheat production step to aquatic acidification and eutrophication, terrestrial ecotoxicity due to the use of chemical products and fertilizers for plant production, and land occupation. The study underlines the contribution of the bioethanol industrial production step to ozone depletion, global warming and acidification due to the use of phosphoric acid, sulfuric acid related to the industrial process and due to the emissions of CO_2, CO, COV and SO_2.

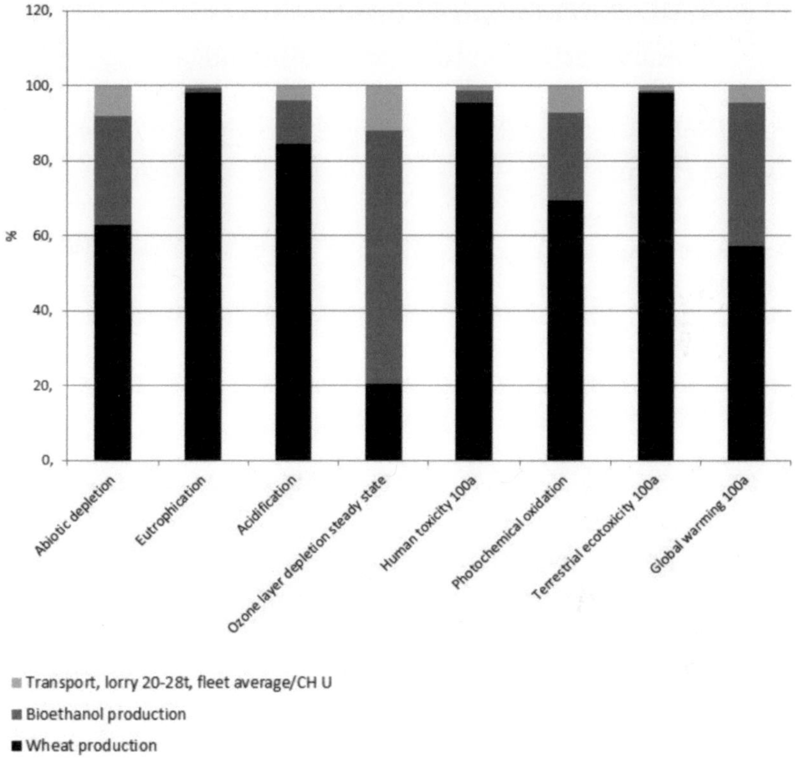

Figure 4.41: Wheat ethanol LCA results.

4.3.4 Impact of bioenergy production and consumption on local air pollution

***Wissal Selmi*[1]**, *Nadège Blond*[1], *Jérémy Stark*[1], *Cyril Pallarès*[2], *Jean-Luc Ponche*[1,3]

[1] Laboratoroire Image, Ville, Environnement (LIVE), Université de Strasbourg/ CNRS, Strasbourg, France.

[2] Association pour la Surveillance de la Pollution Atmosphérique (ASPA), Strasbourg, France.

[3] Institut de Chimie et Procédés pour l'Energie, l'Environnement et la Santé (ICPEES), Université de Strasbourg/ CNRS, Strasbourg, France.

While bioenergy is seen as one of the key options to substitute fossil fuels and contributes to the decrease of greenhouses gases emissions, few studies already claimed that uncontrolled bioenergy consumption may enhance the air pollution at local and regional scale (BIOGAIR 2013; BIOCOMBUST 2015). Since EU air quality thresholds are still often overpassed in some locations, it is important to control that decisions concerning bioenergy will not affect the efforts to reduce air pollution. In order to initiate the assessment of the regional impact of bioenergy production and consumption on air pollution at the scale of the URR, an emission inventory was collected (global inventory discussed in section 2.4) and an air quality modelling study was performed. Since the potential production sites took longer time than expected to be identified (see section 4.2), the air quality modelling study could not be detailed for each options proposed by the economical model. As an alternative, the RA4 scenarios as presented in section 4 were translated in terms of changes in emissions using projections directly discussed with the stakeholders. Due to time restrictions of the project and data availability, the emission scenarios were tested and are presented here only for the Alsace Region for which data on recent bioenergy plants and projects were available. Impacts of these scenarios on the URR air quality are discussed.

4.3.4.1 Curent air pollutant emissions due to biomass utilisation (Alsace Region)

A first analysis of the current energy consumption and associated air pollutant emissions was performed based on the use of energy and emissions inventories provided by ASPA air quality agency (14042904-TD). Information on these inventories is given in section 2.3 and 2.4. Here the analysis is done only on the Alsace Region.

The Alsace energy consumption reached to 231030 TJ in 2010: 38% of this energy is used in industrial sector (manufacturing) followed by residential, road transport and tertiary sectors with respectively 25%, 22% and 13% (the rest concerns other mobile sources and agriculture). The most consumed energy sources are natural gas (32.7%), petroleum product (32%) and electricity (23%). The industrial energy consumption share is 54% for natural gas, 26% for electricity, 7% for petroleum product. The residential sector share is 58% for both electricity and natural gas, 21 % for petroleum product and only 18% for renewable energy.

The total amount of consumed bioenergy reaches 15 608 TJ. The main biomass consumed source is wood biomass (78%) followed by biofuel (21%). The biogas represents only 1%. In 2010 the proportion of sewage sludge and other solid waste is below 1%. The contributions of each type of energy source to the total emissions for each air pollutant in Alsace point out that if many pollutants emissions are issued from fossil energy consumption, wood biomass highly contributes to the total emissions of several pollutants such as CO, PM, HAPs, Benzene Styrene, Toluene (see Figure 2.6 in section 2.4). Other bioenergy consumptions have insignificant effect on the air pollutant emissions.

4.3.4.2 Emission scenarios

Three scenarios were elaborated to assess the impact of an increase of bioenergy production and consumption. These scenarios followed the indications of RA4 scenarios but needed to be quantitatively translated into energy and emissions. Thus, each scenario here illustrates one possible pathway in the general framework of RA4 scenarios but keep the name of the general proposition.

The basecase (BSC) scenario was elaborated from the production and consumption levels of power plants and factories that were already in place and recently implemented in 2010. The Business As Usual (BAU) scenario describes the projection of future energy consumption according to already

planned policies given by the both Regional Climate-Air-Energy plan (SRCAE, 2010) that assume a 26.5% share for renewables of energy production in the final energy consumption expected in 2020. It includes facilities already set up (between 2010 and now) and the ones already decided in different sectors: the actions are summarised in Table 4.16. The ConsRec scenario describes an increase of optimisation of the biomass use, i.e. increase energy efficiency in residential sector and saving of biomass that could be used in other processes. A MaxEx scenario was not possible to be designed without information in time on the locations and the capacities of the new planned production sites, as proposed by the RA2 economical model.

Table 4.16: Drivers of the emission scenarios.

Scenario	Type of action	Measures
BSC	All	Situation and regulation of 2010
BAU	Wood biomass	New plants to produce heat (+52ktep) Substitute fossil energy
	Organic waste	Increase activity of incineration plants (+18ktep) Substitute fossil energy Bioenergy excess transferred to heat network
	Agricultural biomass	Increase energy production (+5 ktep)
	Biofuel	Increase the production of biofuel (+10ktep)
	Efficiency	20% of individual boilers switched to collective boilers and 20% to more efficient individual systems
ConsRec	All	Same as BAU
	Efficiency	50% of individual boilers switched to collective boilers and 30% to more efficient individual systems

Changes in terms of annual air pollutant emissions are presented in Figure 4.42.

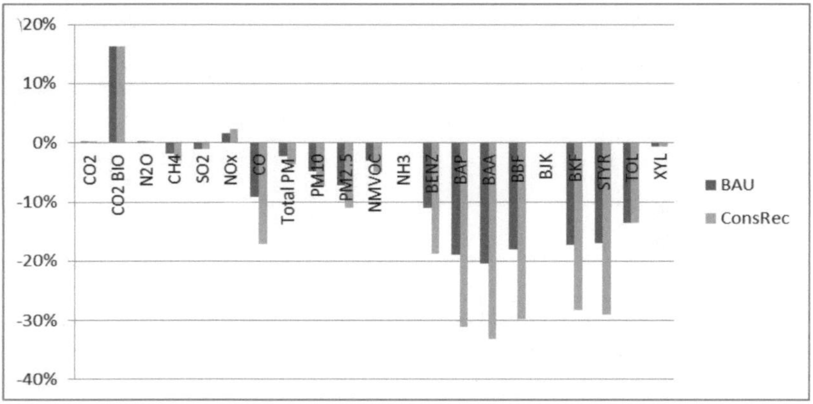

Figure 4.42: Changes in annual air pollutant emissions due to the implementation of BAU and ConsRec scenarios (in %).

It appears that most of air pollutant emissions are reduced due to the implementation of both scenarios BAU and ConsRec. The increase of about 16% of the CO_2 emissions due to biomass consumption (CO_2 BIO) is compensated by the CO_2 absorption by the forest when growing (reduction of the LULUCF sink of around 5%). Other greenhouse gases are not significantly impacted at the scale of the Alsace Region. Only NOx emissions are expected to globally slightly increase (note that NOx is the algebraic sum of NO and NO_2 for both concentrations and emissions): large increases may occur in the residential sector and at the locations of the new bioenergy production plants. Indeed, the temperature rise inside efficient wood-fired boilers improve the efficiency of combustion with the consequence of increasing the NOx and CO_2 emissions, and decreasing CO, CH_4, NMVOC, Benzene and HAPs emissions. While much lower temperatures are usually used for the combustion of agricultural biomass, NOx-emissions can be important due to the limitation of catalyzing effects (Werther et al., 2000). The emissions of the other pollutants are compensated by the substitution of fossil energy. The implementation of collective plants that use filtration devices also helps to reduce particulate matter emissions, benzene, and HAPs.

4.3.4.3 Impacts on air quality

The air quality study was conducted using the air quality modelling system WRF-CHIMERE (Menut et al. 2013), that is a three-dimensional chemistry-transport model developed to simulate gas-phase chemistry and aerosol formation, transport and deposition at regional and urban scales. This system is part of the PREVEST air quality modelling system (ASPA 2014) that is widely applied and regularly validated in the Upper Rhine Region in order to forecast and study air pollution.

The CHIMERE model was first applied to simulate the air pollution in 2010 at European scale with a resolution of $45km^2$, using meteorological inputs from WRF model (Skamarock et al. 2008), emission inventories issued from EMEP (European Monitoring and Evaluation Programme), and boundary conditions issued from LMDz-INCA model (Folberth et al. 2006). The resulting simulations were used as boundary conditions for simulations on two smaller nested grid domains with higher resolutions ($15km^2$ and $3km^2$). The highest resolved grid domain covers Upper-Rhine Region. On this domain, the PREVEST inventories and emission scenarios built in the framework of the 'OUI Biomasse' project were used as inputs to assess the impacts of an increased use of the bioenergy in the Alsace Region by 2020. The emissions out of this area remain unchanged, while it is of course expected that other European regions will also adopt measures on air pollution.

The WRF-CHIMERE system simulates the hourly concentrations of all gaseous air pollutants and aerosols regulated by EU directives (NO_2, ozone O_3, PM2.5, PM10, etc) and many others. The study focuses only on air pollutants for which enough measurements were available for validation. HAPs were thus excluded although they represent important pollutant issued from the biomass burning (specific measurements campaigns are necessary in the Upper Rhine Region to assess the modules allowing the simulations of these species). Figure 4.41 illustrates few results in terms of changes of annual or seasonal average concentrations over the whole URR domain due to the implementation of the measures of BAU and ConsRec scenarios.

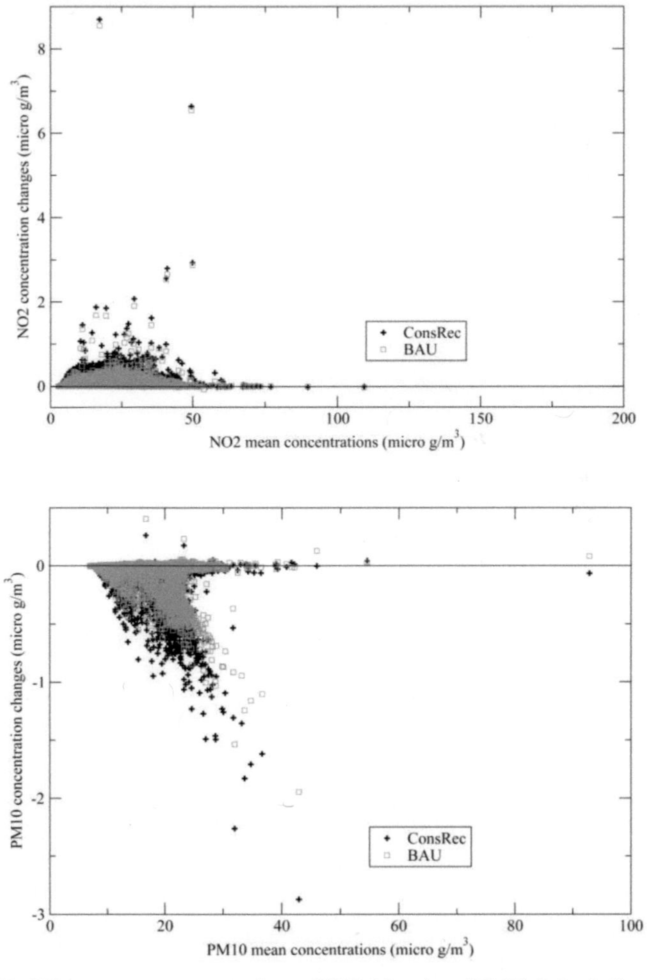

Figure 4.43: Winter average concentrations of NO2 (above) and PM10 (below). Grey squares for BAU. Black crosses for ConsRec.

It was noticed that the measures taken in SRCAE (BAU) and its reinforcement (ConsRec) tend to generally decrease CO, PM10 and PM2.5 concentrations, while local increases on the new plants are noticed (the locations not impacted are usually far away from the Alsace Region where the emission scenarios were tested). The ConsRec scenario has a higher impact than the BAU scenario. Both scenarios lead to general and similar increases of NO_2 concentrations.

Daily and hourly analyses were performed. Figure 4.44 illustrates the simulations of the daily average PM10 and PM2.5 concentrations obtained near Strasbourg city with the ConsRec scenario compared to the base case. It was noted that the ConsRec scenario allows the reductions of the highest PM10 and PM2.5 concentrations (few µg/m3 up to 5 µg/m3), but is limited to avoid PM10 exceedances of the EU threshold (daily average below 50 µg/m3 not to overpass more than 35 days per year).

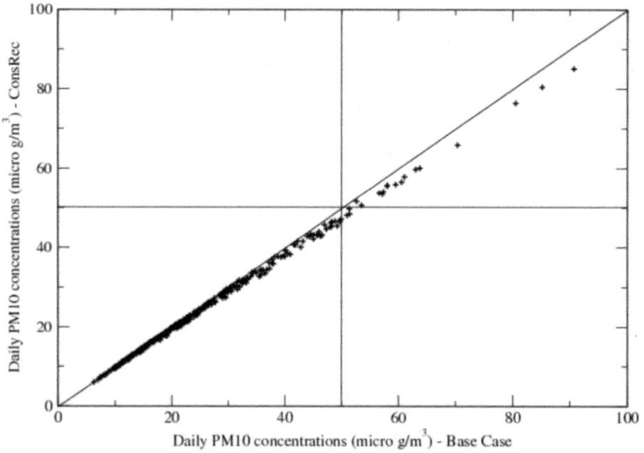

Figure 4.44: Daily mean PM10 and PM2.5 concentrations (in $\mu g/m^3$) simulated with the ConsRec (ordinates) and the Base Case scenarios (abscissa). Line delimits the EU limit values.

To summarise, the production and consumption of bioenergy lead to the emission of new pollutants in the atmosphere but allows reducing the pollution from fossil fuels if it represent an energy substitution. The BAU and

ConsRec scenario showed that the air pollution can be globally reduced at regional scale. The locations of the new plants should be chosen in order to avoid exceedances of EU thresholds at least.

The study has highlighted that the present residential heating sector associated with wood combustion contributes significantly to CO and PM emissions. The use of more efficient household heating systems can globally reduce the air pollution but can also lead to new emissions and increase of NOx concentrations. It is shown that if the BAU and the ConsRec helps to avoid few PM exceedances of EU air quality thresholds, they are limited in their impact (an unrealistic scenario with removing all biomass use shows large PM concentration reductions all over the domain with peak reductions around 40%).

The results are promising but also need to be studied further since the modelling system still has to be improved to better simulate aerosols and also provide information on HAPs. As other models, CHIMERE tends to underestimate PM (e.g. Pirovano at al. 2012). Many efforts need also to be performed to improve more precise and representative emission inventories: spatial distribution of emissions due to residential sector and detailed knowledge about household heat systems should be improved; coherent emission inventories all over URR need to be built (e.g. same reference years inventories over the URR need to be used).

4.3.5 Integrative sustainability assessment to evaluate OUI Biomasse scenarios over URR

Volker Stelzer*, Martin Knapp, Rolf Meyer, Alexandra Pehle*

Institute for Technology Assessment and Systems Analysis (ITAS), Karlsruhe Institute of Technology (KIT), Karlsruhe, Germany.

Sustainability, a term first used in forestry, means carefully managing the available resources to ensure the possibilities for a good life for whole humankind today and in the future.

In order to assess the sustainability impacts of socio-technical developments concerning bioenergy, the widely established Integrative Concept of Sustainable Development (ICoSuD), developed under the lead of the Institute for Technology Assessment and Systems Analysis (ITAS) of the KIT (Kopfmueller et al 2011) was utilised. It takes into account present as well as future generations, and combines thinking on a global scale with actions at a local level. Ensuring human existence and the fundamental needs of all people as well as protecting people's and societies' scope for development and freedom of action are its main objectives. ICoSuD was used to assess the sustainability impacts of the 'OUI Biomasse' scenarios in a systematic way. The main aim was to learn what are the main positive effects and which are the main challenges associated with the different uses of biomass for energy production in the URR.

4.3.5.1 Selection of sustainability rules and indicators

The 15 substantive rules of the ICoSuD were utilised to make sustainability tangible and help implement the objectives mentioned above (Table 4.17).

Table 4.17: The 15 rules of the Integrative Concept of Sustainable Development (Kopfmueller et al. 2011). Rules with relevance for the project 'OUI Biomasse' are marked in grey.

I. Main target: Securing human existence

1. Protecting human health: Nobody may damage the environment through substances or other influences to an extent that will or may harm humans.
2. Ensuring basic needs are met: A minimum level of basic services as well as protection against central risks of life must be guaranteed for all members of society
3. Securing one's livelihood: Every person must be able to secure his livelihood through his own work undertaken voluntarily. This also includes raising children, caring for dependents or community work
4. Equal opportunities of using the environment for everybody: All humans living today and in the future have a right to use nature for themselves within their boundaries. Resources shall be distributed fairly. Nobody may be excluded from their use
5. Ironing out excessive income and wealth imbalances: Extreme differences in income and wealth distribution need to be reduced

II. Main target: Maintaining society's ability to manufacture products or provide services

6. Using renewable resources and energy sources sustainably: Mankind shall not use more from nature than nature is able of its own accord to provide or restore. Any ecosystem important to mankind must have the chance to survive
7. Using non-renewable resources and energy sources sustainably: Non-renewable resources may only be consumed to a limited degree to ensure that future generations will still be able to use them
8. Using the environment without damaging its absorption capacity for harmful substances and waste: Mankind may not release more harmful residues and waste into the environment than it is able to absorb
9. Avoiding unacceptable technical risks. Technical processes with potentially disastrous consequences for mankind and the environment must be avoided
10. Developing property as well as skills and knowledge sustainably: We must leave our descendants an inheritance made up not only of goods but also of skills, competencies, knowledge and know-how

III. Main target: Preserving options for development and action

11. Providing equal opportunities in education, employment, public office, and information. All members of a society must enjoy equal opportunities when it comes to access to education, information, employment, social standing, and political office
12. Ability to take part in social decision-making processes: All members of a society must be empowered to take part in decision-making processes of societal importance
13. Preserving cultural heritage and cultural diversity: The cultural heritage of mankind and its cultural diversity must be preserved
14. Conserving nature and landscape as cultural assets: Especially unique landscapes which have either been created by man or left untouched must be conserved
15. Maintaining social cohesion: Social cohesion must be maintained and strengthened

Depending on whether and how these rules are met, it is possible to analyse a particular measure does or does not make a contribution towards a sustainable development (Kopfmueller, 2011). When it comes to their application, these 15 rules are underpinned and made measurable by indicators. These can be applied specifically on data and background of scientific research, such as different biomass utilisation patterns within the project 'OUI Biomasse'. The outcomes of the resulting final analysis allow decision makers to differentiate between measures that have higher or lower effects on sustainable development.

To start the specific analysis, each of the 15 rules as proposed by Kopfmueller et al. (2011) were subjected to a first estimation of their relevance to the topic of special interest in the context of 'OUI Biomasse': the comparison of sustainability of different biomass utilisation patterns in the Upper Rhine Region. The result of this process was that 9 rules could be selected that are of high relevance to be used in the project (see Table 4.17: labelled in grey). In a next step, suitable indicators to operationalise the selected rules were defined (see Table 4.18) analyzing existing indicator sets applied for the assessment of energy systems and energy related projects (Rösch et al. 2009, Bundesregierung 2014, Löschel et al. 2014, BDI 2013, ZEW 2012, IAEA et al. 2005, Ecoplan, Factor 2001, NEEDS 2006,

CREEA 2014, Destatis 2014, IASS. Adelphi 2012, IASS 2013, Kearney 2012, McKinsey 2014, Ministerium für Umwelt, Klima und Energiewirtschaft Baden-Württemberg, 2013). A main target was to define a low number of indicators ideally fitting to the rules that allow deductions for impacts on regional contexts and that also depict conflicting diverging targets. The applicability of the chosen indicators was verified by comparison to the available data of the project, and then their current status is described (according to Stelzer et al. 2014) via structured fact sheets. The main focus is set on results at the regional level of relevant effects analysed within the single impact studies (see sections from 4.3.1 to 4.3.4) and on biomass production and conversions pathways worked out with the scenarios (see section 4.1).

4.3.5.2 Sustainability analysis of the scenarios

The indicators were applied on the 'OUI Biomasse' scenarios to assess their direction of impacts on sustainability. The results are presented in Table 4.18. The key results are the following:

Table 4.18 Indicators based sustainability assessment of the ‘OUI Biomasse’ scenarios

Indicators	BAU Dev.	BAU SA	MaxEx Dev.	MaxEx SA	Cons-Rec Dev.	Cons-Rec SA
Particle emissions	↓	+	↓↓	++	↓	+
NO_x-emissions	↑	-	↑↑	--	↑	-
NH_3-emissions	→	0	↑	-	→	0
Interruption of electrical power supply	↓	+	↓↓	++	↓	+
Regional employment	↑	+	↑↑	++	↑	+
Regional salaries	↑	+	↑↑	++	↑	+
Non-renewable energy sources	↓	+	↓	+	↓↓	++
Climate relevant emissions	↓	+	↓	+	↓↓	++
Biodiversity	→	0	↓↓	--	↑	+
Soil quality (land use)	→	0	↓↓	--	↑	+
Quality of ground and surface water (land use)	→	0	↓↓	--	↑	+
Distribution of knowledge on biomass utilization	↑	+	↑↑	++	↑	+
Landscape / scenery	→	0	↓	-	→	0

Note: Arrows give the direction of the development (Dev.) of the indicator: arrows pointing up, down, or to the right indicate increase, decrease or stabilization of the indicator. Sustainability assessment (SA) is given using signs: + for positive, 0 for neutral, - for negative effects.

Particles

Small particles can have severe negative health effects, mainly on the respiratory system. The intensification of incineration of wood and biomass in modern heat and power combustion plants in combination with the construction of new district heating networks has the effect that small-scale household fire places are reduced. The emissions of particles in modern plants are much lower than the sum of the older small-scale household fire places. So the amount of particle emissions will be reduced in most parts of the region in all three scenarios. In some local situations and especially in the MaxEx scenario, the reduction of emissions from burning of wood in small-scale household fire places could reduce problematic health situations in wintertime in narrow valleys of some areas, e.g. in the Black Forest and in the Vosges.

NO_x, NH_3 emissions

NO_x has an acidificating and NH_3 a basic effect. So these emissions can have negative impacts on health and on soil and water quality. Another negative effect on water quality is the eutrophicating effect of these substances. They are produced mainly within energy power plants. The production of energy from biogas produces higher rates of NOx and NH_3 than burning of straw and woody biomass. Due to the relative high amount of biogas plants within the MaxEx scenario this vision has to be assessed more negatively than the others. Regarding the health situation the higher NOx emissions decrease the positive effect of the reduction of the particle emissions.

Interruption of electrical power supply

Normally in all three countries the electrical power supply is relatively stable. But sometimes, mainly due to natural hazards, the grid can be interrupted especially in rural area villages but also in small towns or parts of towns which are separated from the main electrical grid. Examples are the destruction of parts of the grid due to landslides, heavy storm, ice or flooding. In this situation local electricity supply based on local resources could help to reduce the negative effects of the electricity cut off.

Regional employment and regional salaries

The scenarios show different impacts in regard to regional economic development potentials. Especially a more intense energetic use of biogenic waste from households, the use of agricultural manure (slurry, dung) and surplus grass from permanent grassland, but also the culture of Miscanthus and short rotation coppice have the potential to stabilise or create new local economic benefits and jobs. The highest positive effect of this is achieved in the MaxEx scenario.

Substitution of non-renewable energy sources and climate relevant emissions

The results of the assessment show that all scenarios have the potential to reduce to some extent the fossil and nuclear energy consumption and the associated emissions of greenhouse gas (GHG). This achievement has multiple reasons: i) the additional use of wood potentials in all three scenarios, ii) the quite minor reduction of the GHG sink potential due to the additional energetic use of wood (see section 4.3.1), iii) the additional cultivation of energy crops in France and Germany (MaxEx scenario), iv) the use of the energy potential in agricultural and domestic residues and wastes, including sewage sludge. In the MaxEx scenario, the positive effect is reduced due to greenhouse gas emissions resulting from ploughing up of permanent grassland. In addition, an expansion of environmental-friendly agricultural production, for example through organic farming, will be restrained. The ConsRec scenario tends to show the best balance regarding these two sustainability indicators.

Biodiversity, soil quality, quality of ground and surface water

Especially in the MaxEx scenario problems with soil quality, biodiversity as well as ground and surface water can emerge. One reason for this is the extension of the cultivation of energy crops in France and Germany and another is the ploughing up of permanent grassland. It is of particular relevance, that a part of the permanent grassland, which would be used for the cultivation of maize, instead, has a relatively high soil quality, a high importance for biodiversity and for the quality of ground water and – indirectly – to surface water. Additionally, there are hints that the cultivation of Miscanthus could have negative effects of the soil pH (see section 4.3.2).

Distribution of knowledge on biomass utilisation

The knowledge about the use of biomass for energy purposes and the skills to implement it will be best developed in the MaxEx scenario due to the highest implementation rate of renewables for energy production.

Landscape / scenery

Forests, grassland and arable land are important parts of historic grown landscapes. On the one hand this has an important function for a great part of the local population for their homeland feelings. On the other hand the beauty of this landscape is an important factor for tourists that connect for example the Black Forest with the wide forests and lovely valleys, mainly grown with grassland. A partly ploughing up of the grassland like in MaxEx could have negative influence to this Indicator.

To summarise, the sustainability method stated that there is not one scenario which could be named as "the sustainable scenario". Every scenario is associated with specific sustainability advantages and disadvantages. Comparing the results obtained with all scenarios, the sustainability indicators consistently show changes in the same direction. These characteristics of the indicators represent the specific advantages and disadvantages of bioenergy production, and, at the same time, indicate relevant areas of action that need to be taken care of to achieve more sustainable energetic biomass uses. The scenarios show only gradual differences. Stronger advantages in the MaxEx scenario are associated with stronger disadvantages. The sustainability assessment of the ConsRec scenario indicates a slight advantage in comparison to the BAU scenario, with a better performance for the indicators "Substitution of non-renewable energy resources" and "Reduction of climate relevant emissions" also as "Biodiversity", "Soil quality" and "Quality of ground and surface water".

4.3.6 Conclusions for section 4.3

Studies were conducted to assess possible impacts of an increase of bioenergy production and consumption on the environment. An integrative sustainability study was also performed to qualitatively evaluate the sustainability of the 'OUI Biomasse' scenarios using environmental, economic and social indicators.

The analysis of the impact of landuse, landuse change and forestry on GHG budget showed that the forest resources of the Upper Rhine region lead to an important sink of GHG (around 17% of the total CO_2 emissions, all activity sectors included). This sink shouldn't be impacted by land use changes as expected by the 'OUI Biomasse' scenarios (forest areas should not be impacted).

Since Miscanthus is seen as a promising energy crop, the impacts of energy crop Miscanthus were detailed through the analysis of soil measurements specifically performed for the 'OUI Biomasse' project. While, it was noted that Miscanthus offer a new potential of CO_2 sequestration in the soil, this CO_2 sequestration mainly concern surface soil, is not permanent, and depends on the season and parcel age. The analysis also showed that the cultivation of Miscanthus leads to a soil acidification.

The Life Cycle Analysis highlights the most damageable processes for health, ecosystem quality, and climate change of several biomasses: wood chips, mixtures of wood pellets or miscanthus with grape marc, and wheat ethanol. The study points out the most important impacts of the timber transport, the woodchips production, the combustion of the biomass in general, the use of pesticides in the wheat production, and the use of the lands. The use of grape marc in the combustion processes tends to enhance the impacts.

The study of the impact of bioenergy production and consumption on regional air pollution was limited to the Alsace Region. It was shown that the scenarios designed to illustrate specific ways among the general 'OUI Biomasse' BAU and ConsRec scenarios allow for a global reduction of main air pollutants, except NO_2 that could show local increases but usually where no exceedances of EU threshold are noticed (i.e. far away from other sources). New regulations are necessary to enhance the replacement of old household heating system by energy efficient ones and locations of new bioenergy production plants has to be decided in order to avoid exceedances of air quality protection indexes.

The detailed previous impact studies were integrated with consideration of social and economic impacts to supply an integrated sustainability study designed to evaluate the sustainability of the 'OUI Biomasse' scenarios. The results show that each scenario is associated with specific sustainability advantages and disadvantages, and none scenario can be selected as the most sustainable.

4.3.7 References section 4.3

Académie d'Agriculture de France (ed.) 1999, *Bilan et gestion des gaz à effet de serre dans l'espace rural. Connaissances tirées du programme AGRIGES et des recherches menées à l'étranger*, Colloque national organisé avec le Ministère de l'aménagement du territoire et de l'environnement, vol. 85, no. 6, pp 1-392.

ASPA 2007, *AtmorhenA, Projet INTERREG III A Rhin supérieur - Projekt INTERREG III A Oberrhein, Système commun d'évaluation et d'information sur la qualité de l'air dans l'espace du Rhin supérieur - Gemeinsames Informations- und Bewertungssystem über die Luftqualität im Oberrheingebiet,* ASPA-07070701-ID, Available from: <http://www.atmo-rhinsuperieur.net/rapport_finalOKencore.pdf>. [7 June 2007].

ASPA 2014, Scores de prévisions de la plate-forme AtmorhenA / PREVEST / ATMO-BW en 2013. [3 October 2014].

BDI (Bund der deutschen Industrie) 2013, *Energiewende-Navigator 2013*, Available from: <http://www.energiewende-richtig.de/download/file/fid/442>.

BIOCOMBUST 2015, *Aspects sanitaires de la production d'énergie à partir de biomasses – gesundheitliche aspekte der energetischen biomassenutzung,* Rapport final, Available from: <www.biocombust.eu>.

Bosansky M 2010, *Evaluation des puits et émissions de CO2 liés à l'utilisation des terres, au changement d'affectation et à la foresterie*, APSA, pp. 1-24.

Bundesregierung 2014, *Zweiter Monitoring-Bericht „Energie der Zukunft"*, Available from: <http://www.bmwi.de/BMWi/Redaktion/PDF/Publikationen/zweiter-monitoring-bericht-energie-der zukunft,property=pdf,bereich=bmwi2012,s-prache=de,rwb=true.pdf >.

Butnar, I, Rodrigo, J, Gazol, CM, Castells, F 2010, 'Life Cycle Asessment of electricity from biomass: case studies of two biocrops in Spain', *Biomass and Bioenergy*, vol. 34, pp. 1780-1788.

Christian, D, Poulton, P, Riche, A, Yates, N 1997, 'The Recovery of 15N-Labelled Fertilizer Applied to Miscanthus X Giganteus', *Biomass and Bioenergy*, vol. 12, no. 1, pp. 21–24.

CITEPA 2012, *Organisation et Méthodes des Inventaires Nationaux des Emissions Atmosphériques en France*, CITEPA, p. 1242.

CREEA 2014, *Compiling and refining environmental and economic accounts*, EU FP7 project, Available from: <http://creea.eu/index.php/documents2>.

Destatis 2014, *Nachhaltige Entwicklung in Deutschland – Indikatorenbericht 2014*, Available from: <https://www.destatis.de/DE/Publikationen/Thematisch/UmweltoekonomischeGesamtrechnun-

gen/Umweltindikatoren/IndikatorenPDF_0230001.pdf?__blob=publicationFile>.

Diets P, 1975, Dichte und Rindengehalt von Industrieholz, *Holz Roh-Werkstoff*, no. 33, pp. 135-141.

Dufossé, K, Drewer, J, Gabrielle, B, Drouet, J-L 2014, 'Effects of a 20-year old Miscanthus giganteus stand and its removal on soil characteristics and greenhouse gas emissions', *Biomass Bioenergy*, vol. 69, pp. 198–210.

Dupouey, J-L, Pignard, G 2000, 'Stocks et flux de carbone dans les forêts françaises', *Revue Forestière Française*, pp. 139-154.

EAB 2012, *Enquête annuelle de production 2011*, Institut national de la statistique et des études économiques.

EEA (European Environment Agency) 2009, *EEA emission inventory guide book 2009 – Uncertainties*, European Environment Agency, p. 26.

ECOBIOM Project 2009, *Compared life cycle assessments of biomass feedstocks at sub-regional scale*, Livrable L7, p.16.

Ecoplan, Factor 2001, *Nachhaltigkeits-Kriterien und Indikatoren für den Energiebereich,* Final report for Swiss Federal Office for Energy BfE, Available from: <http://www.umweltbuero.de/download/data/Nachhaltigkeit._Kriterien_und_Indikatoren_f%FCr_den_Energiebereich_d.pdf>.

FIBOIS 2010, *Quantification de la production et de la consommation locale de produits à destination du bois énergie, dans le cadre de l'observatoire sur les énergies de la CREA*, FIBOIS Alsace.

Foereid, B, de Neergaard, A, Jensen, HH 2004 'Turnover of Organic Matter in a Miscanthus Field: Effect of Time in Miscanthus Cultivation and Inorganic Nitrogen Supply', *Soil Biology and Biochemistry*, vol. 36, no. 7, pp. 1075–1085.

Folberth, GA, Hauglustaine, DA, Lathière, J, Brocheton, F 2006, 'Interactive chemistry in the Laboratoire de Météorologie Dynamique general circulation model: model description and impact analysis of biogenic hydrocarbons on tropospheric chemistry', *Atmospheric Chemistry and Physics*, vol. 6, pp. 2273–2319.

Godard, C, Boissy, J, Gabrielle, B 2012, 'Life Cycle Assessment of local feedstock supply scenarios to compare candidate biomass sources', *GCB Bioenergy*, vol. 5, no.1, pp. 16-29.

Hansen, EM, Christensen, BT, Jensen, LS, Kristensen, K 2004, 'Carbon sequestration in soil be-neath long-term Miscanthus plantations as determined by 13C abundance', *Biomass Bioenergy*, vol. 26, no. 2, pp. 97–105.

Hiederer, R 2013, *Mapping Soil Properties for Europe - Spatial Representation of Soil Database Attributes*, Luxembourg, Publications Office of the European Union, pp. 47.

Howlett, DS, Toma, Y, Wang, H, Sugiyama, s, Yamada, T, Nishiwaki, A, Fernandez, F, Stewart, JR 2013, 'Soil Carbon Source and Accumulation over 12,000years in a Semi-Natural Miscanthus Sinensis Grassland in Southern Japan', *Catena*, vol. 104, pp.127–135.

IAEA, UN DESA, IEA, Eurostat, EEA 2005, *Energy indicators for Sustainable development: Guidelines and methodologies*, Available from: <http://www-pub.iaea.org/MTCD/publications/PDF/Pub1222_web.pdf>.

IASS, Adelphi 2012, *Grundsätzliche Überlegungen und Vorschlag zur Auswahl von Indikatoren zur wissenschaftlichen Begleitung der Energiewende.*

IASS 2013, *Beiträge zur sozialen Bilanzierung der Energiewende*, Available from: < http://www.iass-potsdam.de/sites/default/files/files/ report_beitraege_zur_sozialen_bilanzierung_der_energiewende_0.pdf >.

IFN 2010, *La forêt française, Les résultats issus des campagnes d'inventaire 2005 à 2009, Les résultats pour la région Alsace*, Available from: <http://inventaire-forestier.ign.fr/spip/IMG/pdf/IFN_Publi_2010_Alsace.pdf>. [26 Octobre 2015].

INRA 2002, *Stocker du carbone dans les sols agricoles de France ?*, Institut National de Recherche Agronomique, pp.1-334.

IPCC 2003, *Good Practice Guidance for Land Use, Land-Use Change and Forestry*, Intergovernmental panel on climate change, pp.1-309.

Kaltschmitt, M, Reinhardt, GA, Stelzer, T 1997, 'Life Cycle Analysis of biofuels under different environmental aspects', *Biomass and Bioenergy*, vol. 12, pp. 121-134.

Kearney, AT 2012, *Energiewende-Index WirtschaftsWoche 2012*, Available from: <http://www.wiwo.de/unternehmen/energie/energie-so-weit-ist-die-energiewende/6771830.html>.

Kopfmueller, J 2011, *The integrative sustainability concept of the Helmholtz Association. The "Risk Habitat Megacity" Project as a case of application*. In: Banse, G, Nelson, GL, Parodi, O (eds.): Sustainable development - The cultural perspective. Concepts - aspects - examples. Berlin: edition sigma 2011, pp. 137-149.

Kopfmueller, J, Brandl, V, Joerissen, J, Paetau, M, Banse, G, Coenen, R, Grunwald, A 2011, *Nachhaltige Entwicklung integrativ betrachtet. Konstitutive Elemente, Regeln, Indikatoren*, Berlin, edition sigma.

Liang, S; Xu, M; Zhang, T-Z 2013, Life cycle assessment of biodiesel production in China, *Bioresource Technology*, vol. 129, pp. 72-77.

Lopez, M, Tarelho, L, Ribeiro, I, Monteiro, A, Martins, H, Rafael, S, Miranda, AI, Borrego, C (ed.) 2013, *Impacts of Biomass to Energy chain on air quality and portuguese climate policy (BIOGAIR project),* Universidade de Aveiro, CESAM – Centro de Estudos do Ambiente e do Mar, p. 68.

Löschel, A, Erdmann, G, Staiß, F, Ziesing, H-J 2014, *Expertenkommission zum Monitoring-Prozess „Energie der Zukunft" 2014 Stellungnahme zum zweiten Monitoring-Bericht der Bundesregierung für das Berichtsjahr 2013*, Available from: <http://www.zew.de/de/publikationen/stellungnahme-zum-zweiten-monitoring-bericht-der-bundesregierung-fuer-das-berichtsjahr-2012/?cHash=bf41422266548954b91126c8e2253ef2 >.

Löwe, H, Seufert, G, Raes, F 2000, Comparison of methods used within member states for estimating CO2 emissions and sinks according to UN-FCCC and EU monitoring mechanism: forest and other wooded land, *Biotechnol. Agron. Soc. Environ*, vol. 4, no. 4, pp. 315–319.

McKinsey & Company 2014, *Energiewende-Index*, Available from: <http://www.mckinsey.de/energiewendeindex>.

Ministerium für Umwelt, Klima und Energiewirtschaft Baden-Württemberg 2013, *Monitoring der Energiewende in Baden-Württemberg Schwerpunkt Versorgungssicherheit – Statusbericht 2013*, Stuttgart, Available from: <https://um.baden-wuerttemberg.de/de/presse-service/publikation/did/monitoring-der-energiewende-in-baden-wuerttemberg-schwerpunkt-versorgungssicherheit-1/>.

Menut L, Bessagnet, B, Khvorostyanov, D, Beekmann, M, Blond, N, Colette, A, Coll, I, Curci, G, Foret, G, Hodzic, A, Mailler, S, Meleux, F, Monge, J-L, Pison, I, Siour, G, Turquety, S, Valari, M, Vautard, R, Vivanco MG 2013, 'CHIMERE 2013: a model for regional atmospheric composition modelling', *Geosci. Model Dev.*, vol. 6, pp. 981-1028.

Murphy, F, Devlin, G, McDonnell, K 2013, 'Miscanthus Production and Processing in Ireland: An Analysis of Energy Requirements and Environmental Impacts', *Renewable and Sustainable Energy Reviews* vol. 23, pp. 412–20.

NEEDS 2006, *Establishment of social criteria for energy systems*, Deliverable 2.3, Available from: <EU project. www.needs-project.org/RS2b/RS2b_D2.3.pdf>.

Pirovano, G, Balzarini, A, Bessagnet, B, Emery, C, Kallos, G, Meleux, F, Mitsakou, C, Nopmongcol, U, Riva, G, and Yarwood, G 2012, 'Investigating impacts of chemistry and transport model formulation on model performance at European scale', *Atmos. Environ.*, vol. 53, pp. 93–109.

Poeplau, C, Don, A 2014, 'Soil Carbon Changes under Miscanthus Driven by C 4 Accumulation and C 3 Decomposition - toward a Default Sequestration Function', *GCB Bioenergy*, vol. 6, no. 4, pp. 327–38.

Robertson, GP, Wedin, D, Groffmann, PM, Blair, JM, Holland, EA, Nadelhoffer, KJ, Harris, D 1999, *Soil carbon and nitrogen availability - Nitrogen mineralization, nitrification, and soil respiration potentials*. In: Robertson, P, Coleman, DC, Bledsoe, C, Sollins, P: Standard Soil Method for Long-term Ecological Research (eds.), Oxford University Press, Oxford.

Roger, P, Le Mer, J 2003, 'Les sols: sources et puits de méthane', *Etude et Gestion des Sols*, vol. 10, no. 4, pp. 331-345.

Roth, B, Jones, M, Burke, J, Williams, M 2013, 'The Effects of Land-Use Change from Grassland to Miscanthus X Giganteus on Soil N2O Emissions.' *Land*, vol. 2, no. 3, pp. 437–51.

Rösch, C, Skarka, J, Raab, K, Stelzer, V 2009, 'Energy production from grassland - Assessing the sustainability of different process chains under German conditions', *Biomass & Bioenergy,* vol. 33, no. 4, pp. 689-700.

Rowe, RL, Street, NR, Taylor, G 2009, 'Identifying Potential Environmental Impacts of Large-Scale Deployment of Dedicated Bioenergy Crops in the UK', *Renewable and Sustainable Energy Reviews* vol. 13, no.1, pp. 271–90.

Schéma régional Climat Air Énergie Alsace n.d., Available from: <http://www.alsace.developpement-durable.gouv.fr/IMG/pdf/3-rapport.pdf>. [12 juil. 2010].

Schwartz, D, Namri, M 2002, 'Mapping total organic carbon in the soils of the Congo', *Global and Planetary Change*, vol. 33, pp. 77-93.

Skamarock, W, Klemp, J, Dudhia, J, Gill, D, Barker, D, Wang, W, Powers, J 2008, *A Description of the Advanced Research WRF Version 3*, NCAR

Technical Note, Boulder, Colorado, USA, NCAR/TN–475+STR, Available from: <http://www2.mmm.ucar.edu/wrf/users/docs/arw_v3.pdf>.

Smeets, EMW, Lewandowski, IM, Faaij, APC 2009, 'The Economical and Environmental Performance of Miscanthus and Switchgrass Production and Supply Chains in a European Setting', *Renewable and Sustainable Energy Reviews*, vol. 13, pp. 1230–1245.

Stelzer, V, Quintero, A, Vargas, L, Paredes, G, Simon, S, Nienhaus, K, Kopfmüller, J 2014, *Indicator based sustainability analysis of future energy situation of Santiago de Chile*. In: Awrejcewicz, J, Shitikova, M, Niola, V, Panagopoulos, T, Wenzel, W, Gorunescu, F, Horova, I, Korobeinikov, A (eds.): Energy, Environment, Biology and Biomedicine. Prague, Czech Republic: 2014, pp. 24-27.

Sungdo, K, Dale, BE 2005, 'Life Cycle Assessment of various cropping systems utilized for producing biofuels: Bioethanol and biodiesel', *Biomass and Bioenergy*, vol. 29, pp. 426-439.

Tobin, B, Nieuwenhuis, M 2005, 'Biomass expansion factors for Sitka spruce (picea sitchensis (Bong.) Carr.) in Ireland', *European Journal of Forest Research*, vol. 126, pp. 189-196.

TRION 2012, *Chiffres clés sur l'énergie du Rhin supérieur*, Réseau énergie-climat, Région Métropolitaine Trinationale du Rhin Supérieur, p. 24.

van der Hilst, F, Lesschen, JP, van Dam, JMC, Riksen, M, Verweij, PA , Sanders, JPM, Faaij, APC 2012, 'Spatial Variation of Environmental Impacts of Regional Biomass Chains', *Renewable and Sustainable Energy Reviews*, vol. 16, no. 4, pp. 2053–2069.

Werther, J, Saenger, M, Hartge, E-U, Ogada, T, Siagi, Z 2000, 'Combustion of agricultural residues', *Prog. Energy Combust. Sci.*, vol. 26, pp. 1–27.

Yang, J, Zhang, P 2011, 'Assessment Methods of Carbon Dioxide Emitted from Bioenergy Utilization', *Renewable and Sustainable Energy Reviews*, vol. 15, no. 6, pp. 2684–2689.

Zibilske, LM 1994, 'Carbon mineralization', *Soil Science Society of America*, Methods of Soil Analysis, pp. 835–863, Madison, Wisconsin, USA.

ZEW (Zentrum für europäische Wirtschaftsforschung) 2012, *Indikatoren für die Energiepolitische Zielerreichung*, Available from: <http://ftp.zew.de/pub/zew-docs/gutachten/ZEW_Indikatorenbericht_2012.pdf>.

Zimmermann, J, Dondini, M, Jones, MB 2013, 'Assessing the impacts of the establishment of Miscanthus on soil organic carbon on two contrasting land-use types in Ireland', *European Journal of Soil Science*, vol. 64, no. 6, pp. 747–756.

5 Roadmap towards a more sustainable use of biomass in the URR

***Olga Steiger*[1], *Nadège Blond*[2], *Christoph Hugi*[1], *Rolf Meyer*[3]**

[1] Institut für Ecopreneurship (IEC), Hochschule für Life Sciences, Fachhochschule Nordwestschweiz, Muttenz, Switzerland.

[2] Laboratoire Image, Ville, Environnement (LIVE), Université de Strasbourg/ CNRS, Strasbourg, France.

[3] Institute for Technology Assessment and Systems Analysis (ITAS), Karlsruhe Institute of Technology (KIT), Karlsruhe, Germany.

5.1 Options for action for a sustainable energetic biomass utilisation in the URR

Based on the current situation described in section 2 and the scenarios described in section 4, following five clusters of action options have been identified within the project. The first cluster of action option, "ensuring sustainability of energy production" is at the same time a premise, which is valid for all following action options.

- Ensuring sustainability of bioenergy production
- Improved efficiency of existing bioenergy conversion pathways and bioenergy uses
- Restricted mobilisation of additional biomass for bioenergy production
- Strengthening of regional innovation
- Improve database on bioenergy production in the URR as a baseline for decision making

5.1.1 Ensuring sustainability of bioenergy production

Sustainability assessments should be implemented into biomass use planning

Varying pathways of bioenergy production and use of biomass are increasingly under discussion with regard to their sustainability impacts. Here contradictory positions have developed, e.g. on the topic of biofuels. Thus, acceptancy of bioenergy production cannot be generally assumed with stakeholders and citizens affected. Involving them into the formulation of regional energy strategies could be a helpful approach. To make different arguments transparent and debatable, sustainability assessments should be conducted while further developing regional strategies and concrete projects. In doing so, various criteria and dimensions of sustainability are to be taken into consideration in an integrative way. Relevant aspects are superior, like for example the long-term preservation of soil fertility in the region, or project specific, like biomass transportation distances. These should be minimised while planning and managing bioenergy plants, in order to reduce the corresponding emissions and other negative impacts like noise. This is crucial to ensure the key argument of ecologic advantages of the regional biomass use as energy source.

Responsible actors: Plant operators, biomass supplier, energy supplier, project developer, regional policy maker, regional administration, civil society organisations

5.1.2 Improved efficiency of existing bioenergy conversion pathways and bioenergy uses

The efficiency of established energy conversion pathways should be improved

Energetic use of biomass has already partly reached a high percentage of existing potentials, especially in the wood sector. Therefore, the main focus should be on efficiency improvement of existing bioenergy production to produce more bioenergy from the restricted regional biomass resources. Investments into upgrading and replacement can be supported with specific incentives in order to increase the efficiency and ensure sustainability. Additionally, alternative concepts, such as wood and biogas use for industrial high temperature processes should be evaluated in respect to their contribution to increase the sustainability of bioenergy use.

Responsible actors: Pant operators, project developers, regional policy makers

Improvement of heat use in bioenergy plants

Bioenergy plants in general and biogas plants in particular often still do not use all of the heat produced by their CHP units. Therefore, options for optimal heat use or alternatively upgrade of produced biogas to biomethane with injecting into a gas grid should be evaluated. In the case of biomethane, an important advantage is that the heat use is independent from the biogas production and the possibility to use biogas as fuel. The promotion of biomethane production should be more integrated into the promotion of electricity and heat production from biomass.

Responsible actors: Plant operators, technology providers, energy suppliers, grid operators.

Upgrade of small-scale fire places should modern technologies for burning processes or replacement with district heating networks

Negative effects of the increasing energetic use of wood in small-scale fire places could be reduced by optimisation with modern technology for burning processes or by installation of filter flue gas. Another path to go is the construction or further development of local heating networks with central, optimises heat production in densely populated areas. In this case, also biomass use efficiency would be improved. However, this solution is only applicable within settlements with a certain heat demand by private and business building.

Responsible actors: Plant operators, technology providers, local authorities

Controlled use of demolition wood (scrap wood)

The data basis of demolition wood usage was small in this project and should be improved. It is expected that the potential in the long run must be in the same order of magnitude as the material use of wood in the region and should therefore be energetically optimally exploited, i.e. avoiding improper burning, long transportation distances and ensuring maximum use of energy.

Responsible actors: Waste disposal companies, waste incineration plants

Wet biogenic fraction of household waste should be redirected from incineration or composting to digestion before composting

In addition to the established conversion pathways of waste incineration and composting, industrial biogas plants are interested in the same sources of biomass. More separate collection of organic waste and digestion before composting could be a preferable long-term alternative from a cascade use and multiple valorisation perspective. Contaminations in these biomass streams may however result in required post treatment similar to plans for wastewater sludge, i.e. mono-incineration and nutrients recovery. Provided a successful technological development, these biomass fractions could be used also by other valorisation technologies in future, for example such as hydrothermal carbonization.

Responsible actors: Local policy makers and administration, operators of waste incineration plants, compost works and biogas plants.

5.1.3 Restricted mobilisation of additional biomass for bioenergy provision

The technical potentials of agricultural residues and manure for biogas production should be better exploited

Although there are technical potentials for energetic use of agricultural residues and manure, regional distribution and the average size of farms require biogas concepts which are efficient also at a small scale. Beside innovation in small-scale profitable biogas plants, openness of farmers to engage in biogas production should be respected and specific regional incentives and support should be considered.

Responsible actors: Technology providers, biogas plant operators, farmers

Increased energetic use of wood residues should be restricted by ecological requirements

Under the premise of priority for material uses, an extension of energy wood production is not desirable. Residues of harvested stem and industry wood are already partly used for energy production. The regional potential of unexploited wood residues is difficult to estimate, but restricted potentials are available, depending on local conditions of e.g. nutrition recycling and nature protection. Obligations from forest certification have also to be taken into account.

Responsible actors: Forest authorities, private forest owners, biomass suppliers, bioenergy plant operators

Competition between different biomass conversion pathways is growing and should to be actively managed and coordinated

Most of biomass resources can be employed in different uses and bioenergy conversion pathways. Strong competition exists already for many agricultural crop products and forestry wood. Also organic residues from agriculture, organic waste from industries and the organic fractions of household waste come under increasing competition. Such competitions should be considered for a successful project planning. A local assessment of economic, environmental and social benefits of the different bioenergy conversion pathway options and their combinations could be helpful to optimise energetic biomass use. These activities require a more intensive involvement of relevant stakeholders and an open dialogue. Networks and longer-term supply contracts for a stable local biomass supply can be achieved by engaging biomass suppliers in bioenergy projects as shareholders.

Responsible actors: Local and regional policy maker, local/regional administration, biomass provider, project developer, bioenergy plant operators, civil society organisations

Strongly limited chances for additional cultivation of energy crops in the URR

Future reductions or abolishment of quota for biofuel use in the EU will have only a small effect on the land use in the URR. Therefore, also only restricted arable land and energy crop production for new regional energetic uses could be expected. Without strong changes in the EU and/or national renewable energy policies, new energy crop cultivations will remain restricted to a small number of specific situations (e.g. short rotation coppice on marginal land) but should be exploited.

Responsible actors: Farmers, plant operators

5.1.4 Strengthening regional innovation

Support for regional innovation and pilots should be increased

Limited regional potentials of biomass require corresponding innovative solutions which work well with restricted amount of biomass per bioenergy plant. Changed financial support frameworks (e.g. EEG amendment in Germany) should motivate technology providers and project developers to work on innovative solutions which deal successfully with the reduced support and under regional conditions. Additionally, different legislative frameworks in the three sub-regions hamper a cross-border application of bioenergy technologies and plant concepts. Target-oriented regional activities and programs for research and innovation could at least help to overcome these barriers.

Responsible actors: Regional policy maker, research institutions, technology providers, project developer

Programs for local bioenergy concepts and implementations should be continued and spread across the boarders

The federal program of bioenergy villages (Bioenergiedörfer) in Germany has been successful in promoting the use of regional biomass resources for bioenergy production. A bioenergy village covers a large part of its electricity demand and heat requirements by local production through the use of mainly regionally supplied biomass. Energy city is a similar label which is widely used in Switzerland for sustainable energy management on the municipal level (Energiestadt) and promotes the use of renewable energy in general. Two possible actions are seen in the context of these successful programs: on the one hand, regional actors should advocate for a continuation or introduction by the national government, on the other hand, local initiatives should be supported in making use of these opportunities.

Responsible actors: Regional policy makers

5.1.5 Improve database on bioenergy production in the URR as a baseline for decision making

Data collection and availability on biomass and bioenergy production in the URR should be improved

Currently data on biomass production and potentials is often available only in a heterogeneous form or is incomplete or not directly comparable. Often different categories of data are used or data is aggregated at different regional and product group levels. These problems arise for all examined biomass sources: forest area and wood harvest, agricultural land use and production, including residues and manure, as well as biogenic wastes. In consequence, data collection, evaluation and accessibility should be improved in the region. This would require more exchange and alignment of statistical approaches by the responsible administrations.

Responsible actors: URR offices, regional administration, national and regional statistics departments

Bioenergy production and biomass supply pathways in the region should be regularly monitored

Information on the current bioenergy production and associated biomass supply chains is incomplete and differs depending on bioenergy conversion pathway and sub-region. Also, the understanding and differentiation of bioenergy producing plants is different in the three countries. Important basic data for bioenergy strategy development and project planning are in consequence missing. Setting up a periodic monitoring of bioenergy production development at the regional level would improve the decision making in the region. Once again, precondition for such a monitoring is a common approach on data collection, evaluation and presentation along the biomass conversion pathways.

Responsible actors: URR offices, regional administration, national and regional statistics departments

Information exchange and coordination should be enhanced in regard to different sub-regional bioenergy strategies and the exploitation of optimisation potentials

Most of legislation concerning biomass production and use are set up at national level in the corresponding strategies and regulations on renewable energy and differ substantially between the three countries. Given the regional character of biomass, in the long term a cross-country strategy for energetic biomass use should be considered. This would require an increased exchange of information and coordination between the involved countries. At the same time, this would provide a possibility to learn in mutual learning processes form success stories and failures.

Responsible actors: National and regional politics, decision makers, associations

6 Glossary

Agricultural residues are by-products of agricultural production with primary residues arising from the cultivation and harvesting of plants like straw and other plant remains. Secondary residues stem from the processing of the harvested plants in food production or other processing facilities, typical residues can be husks, kernels or peels for example. Manure deriving from livestock breeding is a separate category of agricultural waste.

Bioenergy is energy derived from the conversion of biomass where biomass may be used directly as fuel, or processed into liquids and gases.[4]

Biogas plants digest energy crops, manure and other organic residues to biogas, which is converted to electricity and heat by combined heat and power plants. Some biogas plants upgrade the biogas processed to biomethane, which, fed into the gas grid, can be used as fuel.

Biomass is defined as all organic material which has been produced directly or indirectly through photosynthesis processes and which has not been changed by geological processes (like fossil fuels such as oil, coal or natural gas). Main considered biomass types for this report are forestry products, agricultural crops in general or energy crops more specifically, agricultural residues including manure, and organic waste fractions and sludge generated by wastewater treatment plants.

Energy crops are crops specifically grown for energetic biomass use. Energy crops on agricultural or marginal land consist in either annual or perennial crops and can be divided in five different categories: 1. Oil crops

[4] IEA: https://www.iea.org/topics/renewables/subtopics/bioenergy/

like sunflower, rape, soy and oil palm, 2. Sugar crops like sugar beet, 3. Starch crops like corn, wheat and barley, 4. Woody crops deriving from short rotation coppice and 5. Grassy crops like Miscanthus.

Forestry biomass can be derived from managed forests, short rotation plantations on forest lands and trees in settlement or infrastructural areas. The used materials can either be stem wood, i.e. whole trees or delimbed stem wood that comes from (pre-) commercial thinnings and final fallings or other forestry residues. Primary forestry residues consist in stumps and leftover residues from logging, such as branches, twigs and leaves. Secondary forestry residues arise by processing wood, typical examples of these would be sawdust, bark, cutter chips and black liquor.

Organic waste consists of residues stemming from industry and trade activities and from households in general. This includes all forms of biodegradable waste, like municipal waste, wood from construction and old furniture, sewage gas and sludge and landfill gas.

Sewage gas plants are operated at wastewater treatment plants and use digestion technology to process mainly sewage sludge to produce electricity and heat by combined heat and power plants, which is mainly consumed by wastewater treatment plants.

Waste incineration plants use the waste-to-energy process to generate energy in form of electricity and/or heat from the incineration of waste. In France, Germany and Switzerland 50% of electricity generated by waste incineration plants is accounted as renewable.

Wood heating and power plants generate heat and electricity in combined heat and power plants.

Wood furnaces are automatic wood combustion plants, which produce heat only through combustion of pellets, chips, logs and wood residues.

PRODUKTION UND ENERGIE

Karlsruher Institut für Technologie (KIT)
Institut für Industriebetriebslehre und Industrielle Produktion
Deutsch-Französisches Institut für Umweltforschung

ISSN 2194-2404

Die Bände sind unter www.ksp.kit.edu als PDF frei verfügbar oder als Druckausgabe zu bestellen.

Band 1 **National Integrated Assessment Modelling zur Bewertung umweltpolitischer Instrumente.**
Entwicklung des otello-Modellsystems und dessen Anwendung auf die Bundesrepublik Deutschland. 2012
ISBN 978-3-86644-853-7

Band 2 **Erhöhung der Energie- und Ressourceneffizienz und Reduzierung der Treibhausgasemissionen in der Eisen-, Stahl- und Zinkindustrie (ERESTRE).** 2013
ISBN 978-3-86644-857-5

Band 3 Frederik Trippe
Techno-ökonomische Bewertung alternativer Verfahrenskonfigurationen zur Herstellung von Biomass-to-Liquid (BtL) Kraftstoffen und Chemikalien. 2013
ISBN 978-3-7315-0031-5

Band 4 Dogan Keles
Uncertainties in energy markets and their consideration in energy storage evaluation. 2013
ISBN 978-3-7315-0046-9

Band 5 Heidi Ursula Heinrichs
Analyse der langfristigen Auswirkungen von Elektromobilität auf das deutsche Energiesystem im europäisschen Energieverbund. 2013
ISBN 978-3-7315-0131-2

Band 6 Julian Stengel
Akteursbasierte Simulation der energetischen Modernisierung des Wohngebäudebestands in Deutschland. 2014
ISBN 978-3-7315-0236-4

Band 7 Sonja Babrowski
Bedarf und Verteilung elektrischer Tagesspeicher im zukünftigen deutschen Energiesystem. 2015
ISBN 978-3-7315-0306-4

Band 8 Marius Wunder
Integration neuer Technologien der Bitumenkalthandhabung in die Versorgungskette. 2015
ISBN 978-3-7315-0319-4

Band 9 Felix Teufel
Speicherbedarf und dessen Auswirkungen auf die Energiewirtschaft bei Umsetzung der politischen Ziele zur Energiewende. 2015
ISBN 978-3-7315-0341-5

Band 10 D. Keles, L. Renz, A. Bublitz, F. Zimmermann, M. Genoese, W. Fichtner, H. Höfling, F. Sensfuß, J. Winkler
Zukunftsfähige Designoptionen für den deutschen Strommarkt: Ein Vergleich des Energy-only-Marktes mit Kapazitätsmärkten. 2016
ISBN 978-3-7315-0453-5

Band 11 Patrick Breun
Ein Ansatz zur Bewertung klimapolitischer Instrumente am Beispiel der Metallerzeugung und -verarbeitung. 2016
ISBN 978-3-7315-0494-8

Band 12 P. Ringler, H. Schermeyer, M. Ruppert, M. Hayn, V. Bertsch, D. Keles, W. Fichtner
Decentralized Energy Systems, Market Integration, Optimization. 2016
ISBN 978-3-7315-0505-1

Band 13 Marian Hayn
Modellgestützte Analyse neuer Stromtarife für Haushalte unter Berücksichtigung bedarfsorientierter Versorgungssicherheitsniveaus. 2016
ISBN 978-3-7315-0499-3

Band 14 Frank Schätter
Decision support system for a reactive management of disaster-caused supply chain disturbances. 2016
ISBN 978-3-7315-0530-3

Band 15 Robert Kunze
Techno-ökonomische Planung energetischer Wohngebäudemodernisierungen: Ein gemischt-ganzzahliges lineares Optimierungsmodell auf Basis einer vollständigen Finanzplanung. 2016
ISBN 978-3-7315-0531-0

Band 16 A. Kühlen, J. Stengel, R. Volk, F. Schultmann, M. Reinhardt, H. Schlick, S. Haghsheno, A. Mettke, S. Asmus, S. Schmidt, J. Harzheim
ISA: Immissionsschutz beim Abbruch - Minimierung von Umweltbelastungen (Lärm, Staub, Erschütterungen) beim Abbruch von Hoch-/Tiefbauten und Schaffung hochwertiger Recyclingmöglichkeiten für Materialien aus Gebäudeabbruch. 2017
ISBN 978-3-7315-0534-1

Band 17 Konrad Zimmer
Entscheidungsunterstützung zur Auswahl und Steuerung von Lieferanten und Lieferketten unter Berücksichtigung von Nachhaltigkeitsaspekten. 2016
ISBN 978-3-7315-0537-2

Band 18 Kira Schumacher, Wolf Fichtner and Frank Schultmann (Eds.)
Innovations for sustainable biomass utilisation in the Upper Rhine Region. 2017
ISBN 978-3-7315-0423-8